Heidelberg
Science
Library

Juan G. Roederer

Introduction to the Physics and Psychophysics of Music

Springer-Verlag
New York
Heidelberg
Berlin

SECOND EDITION
Second Reprint, Corrected

Juan G. Roederer
Geophysical Institute
University of Alaska
Fairbanks, Alaska, 99701

Library of Congress Cataloging in Publication Data

Roederer, Juan G.   1929–
   Introduction to the physics and psychophysics of music.

   (Heidelberg science library: v. 16)
   Includes bibliographies and index.
   1. Music—Acoustics and psychophysics. I. Title.   II. Series.
ML3805.R74   1975   781'.22   75-2313

Cover illustration by J. G. Roederer. Composed using statistical
distribution histograms of time intervals between neural impulses
detected by microelectrodes implanted in auditory nerve fibers
of cats and squirrel-monkeys (Section 2.9); and oscillograms of
second order beats of a mistuned octave (Section 2.6). The
author is indebted to Dr. J. E. Rose, University of Wisconsin, for
his permission to use a collection of histograms for this purpose
(see reference Rose et al. 1969).

Printed in the United States of America.

ISBN 0-387-90116-7   Springer-Verlag   New York

ISBN 3-540-90116-7   Springer-Verlag   Berlin Heidelberg

9 8 7 6 5 4 3

Dedicated to the memory of
my dear father

ABOUT THE AUTHOR

The author, Professor Juan G. Roederer, is a space scientist of international reputation. Italian born, he spent most of his life in Buenos Aires, Argentina, where he received his Ph.D. in physics at the local university. He conducted research for several years at the Max Planck Institute for Physics in Göttingen, Germany, and at NASA's Goddard Space Flight Center in Greenbelt, Maryland. For ten years he was Professor at the University of Buenos Aires and a member of its Directive Council. In 1966 he and his family decided to emigrate to the United States, where he took the position of Professor of Physics at the University of Denver, Colorado. Since 1977 he is director of the Geophysical Institute of the University of Alaska in Fairbanks. Author of several books and over 100 scientific articles, Professor Roederer is President of the International Association of Geomagnetism and Aeronomy, Chairman of the International Magnetospheric Study, and a member of several committees and panels of the National Academy of Sciences and other national and international scientific organizations.

Professor Roederer's close and active relationship with music—he studied organ with Héctor Zeoli in Buenos Aires and Hans Jendis in Göttingen—prompted him, several years ago, to organize a course on Physics of Music and Psychoacoustics at the University of Denver, of which the present book is an offspring. The teaching of this course has switched his interest part-time to the "inner space" of the brain functions, in particular to the study of neuropsychological mechanisms responsible for the processing of acoustical information.

Preface to the First Edition

This book deals with the physical systems and psychophysical processes that intervene in what we broadly call "music." We shall analyze what objective, physical properties of sound patterns are associated with what subjective, psychological sensations of music. We shall describe how these sound patterns are actually produced in musical instruments, how they propagate through the environment, and how they are detected by the ear and interpreted in the brain. We shall do all this by using the physicist's language and his method of thought and analysis—without, however, using complicated mathematics (this, of course, will necessarily impose serious limitations on our presentation). Although no previous knowledge of physics is required, it is assumed that the reader is familiar with music, in particular with musical notation, musical scales and intervals, that he has at least some basic ideas about musical instruments, and that he has experienced typical musical "sensations."

Until about 25 years ago, little attention had been paid to the role of the brain, i.e., the central nervous system, in the actual perception, identification, and evaluation of musical sounds. The highly "mechanicistic" approach of 19th-century researchers, notably the great von Helmholz (1863), persisted well into the first half of this century. In this approach the perception of musical tones was regarded mainly as a result of the conversion of certain well-defined sound wave properties (frequency, intensity, and spectrum) into more or less well-defined groups of neural signals (coding information on pitch, loudness, and timbre, respectively). Today we know that the central nervous system does play a far more active role; a role so essential that without it the perception of even such basic attributes as the pitch of musical tones would be impossible. This is why in this book we shall refer a great deal to the discipline of psycho-physics, which in a broad sense tries to establish in a quantitative form the causal relationship between the "physical" input from our senses and the psychological sensations and physiological reactions evoked in our mind and body, respectively. Actually, we shall try to weave a rather close mesh between physics and psychophysics—or, more precisely, psychoacoustics. After all, they appear naturally interwoven in music itself: not only pitch, loudness and timbre are a product of physical and psychoacoustical processes, but so are the sensations related to consonance and dissonance, tonic dominance, trills and ornamentation, vibrato, phrasing, beats, tone attack, duration and decay, rhythm, and so on.

Many books on physics of music or musical acoustics are readily available. An up-to-date text is the treatise of John Backus (1969). No book on psychoacoustics is available at the elementary level, though. Several review articles on pertinent topics can be found in Tobias (1970) and in Plomp and Smoorenburg (1970). A comprehensive discussion is given in Flanagan's book on speech (1972). And, of course, there is the classical treatise of von Békésy (1960). A comprehensive up-to-date analysis of general brain processes can be found in Sommerhoff (1974); musical psychology is discussed in classical terms in Lundin (1967). A periodical publishing original research articles on physics of music and psychoacoustics is the *Journal of the Acoustical Society of America*. The purpose of this book is not to duplicate but to synthesize existing literature. The blending of physics and psychoacoustics has been the main guideline followed by the author.

One of the most painful parts of writing a book is deciding what topics should be left out, or grossly neglected, in view of the stringent limitations of space. No matter what the author does, there will always be someone bitterly complaining about this or that omission. Let us list here some of the subjects that have been neglected or omitted—without venturing to attempt a justification. In the discussion of the generation of musical tones mainly basic mechanisms are analyzed, to the detriment of the presentation of concrete musical situations. The human voice has been all but left out and so have discussions of inharmonic tones (percussion instruments) and electronic tone generation; computer-generated music is not even mentioned. On the psychoacoustical side, only the perception of single or multiple *sinusoidal* tones is analyzed, with no word on noise-band or pulse stimuli experiments. Cochlear hydrodynamics and electrophysiology, as well as a description of the neuro-anatomy of the auditory tract have had to be omitted, too. And there is practically nothing on rhythm, almost nothing on stereo perception, and very little on historical development. Finally, in literature references, the emphasis is on psychoacoustical subjects. Priority was given to the quotation of reviews in sources of more widespread availability to the general public; detailed references of original articles can be found in most of the quoted reviews.

This book is an offspring of a syllabus published by the University of Denver for the students in the Physics of Music course, which was introduced at the university in the fall of 1970. In addition to regular classwork, these students are required to perform a series of acoustical and psycho-acoustical experiments in a modest laboratory. Conducting such experiments, some of which will be described, is essential for a clear comprehension of the principal physical and psychoacoustical concepts involved. Unfortunately for the general reader, they often require electronic equipment that is not readily available. We ask that the readers trust our description of the experiments and believe that they really do turn out the way we say they do. Whenever possible, we shall indicate how a given experiment can be performed with the aid of ordinary musical equipment.

The author is grateful to Professor A. H. Benade of Case Western Reserve University (Cleveland), and to Professor R. T. Schumacher of Carnegie–Mellon University (Pittsburgh) for helpful comments and criticism. Sections 4.5 and 6 of this book are based on recent, still mostly unpublished, work done by Professor Benade and coworkers. The author also acknowledges with gratitude the efficient secretarial work of Norma Lanier, the expert technical drawing of James Haworth and photographic work of David Clint, and the careful revision of the manuscript by the author's wife, Beatriz.

Preface to the Second Edition

Only one year has passed since the publication of the first edition of this book. It is a gratifying sign that the first printing has sold out after such a short time. A lot has happened since the submission of the original manuscript (mid-1972), especially in the field of psychoacoustics. Also, much thoughtful feedback has been received from colleagues and students who have used this book in class.

The main revisions in this second edition take into account important recent developments in the understanding of complex tone pitch perception (Sections 2.7, 2.8, 2.9, and 4.8). Look at the letters of the title page. Do you see contours which actually are not physically present? Part of the revisions of this second edition consists of the introduction of an auditory analog to this visual phenomenon, as a possible explanation of the pitch perception of complex tones (see references Terhardt 1974, Goldstein 1974, and Wightman 1973). Related to this, some new ideas on consonance and dissonance have also been incorporated (Section 5.2). Section 2.9 is new and contains a brief description of the principal information channels in the auditory pathway. The specialization of cerebral hemispheres with regard to speech and music processing is the subject of new Section 5.4. In the "physical" sections, only a few corrections or clarifications have been made. We believe that now this book is even more "interdisciplinary" than was its first edition.

Many of the comments received from readers call for more figures and more truly musical examples, and suggest the inclusion of problems and questions for use in class. Unfortunately, at this time it is impossible to expand this monograph into a full-fledged textbook. If encouragement continues, I may do so in the future. A first small step in this direction is the addition of Appendix III, which deals with some aspects of teaching the subject.

Juan G. Roederer

Denver, January 1975

Preface to the Second Edition, Second Reprint, Corrected

In 1977 my family moved to Alaska where I assumed the position of director of the Geophysical Institute of the University of Alaska. This is a world-renowned research institute and my job is extremely challenging—but a high price had to be paid: giving up entirely teaching and research in musical acoustics and psychoacoustics! No time was left to keep up in detail with the steady progress made in recent years in this field, and concomitantly no time was left to prepare a thoroughly revised or expanded third edition. I take comfort in the impression that the present version cannot be that much in need for change: it has been translated into German (German edition 1977) and is being translated into Japanese (to appear later this year). I even have discovered a book, recently published, that has borrowed most heavily from mine in organization, text and figures! These are all most gratifying indicators, especially in view of the fact that the topic in question is—was—more of a hobby than a full-time occupation.

In spite of all this, I still wish to maintain contact with teachers, investigators and students of musical acoustics and psychoacoustics. Please do write me with any questions or criticism you may have, please keep in touch!

The reader interested in expanding his or her knowledge beyond what is contained in this edition is strongly encouraged to turn to the following new books: Arthur Benade's treatise *Fundamentals of Musical Acoustics* (Oxford University Press, New York, London, Toronto, 1976); Jürgen Meyer's *Acoustics and the Performance of Music* (Verlag Das Musikinstrument, Frankfurt a/M, Germany, 1978), and Reinier Plomp's *Aspects of Tone Sensation* (Academic Press, London, New York, San Francisco, 1976).

Juan G. Roederer

Fairbanks, February 1979

Contents

Music, Physics, and Psychophysics: An Interdisciplinary Approach 1

1.1 The intervening physical systems *1* / **1.2** Characteristic attributes of musical sounds *3* / **1.3** The time element in music *5* / **1.4** Physics and psychophysics *7* / **1.5** What is music? *11*

Sound Vibrations, Pure Tones, and the Perception of Pitch 13

2.1 Motion and vibration *13* / **2.2** Simple harmonic motion *17* / **2.3** Acoustical vibrations and pure tone sensations *18* / **2.4** Superposition of pure tones: first-order beats and the critical band *25* / **2.5** Other first-order effects: combination tones and aural harmonics *33* / **2.6** Second-order effects: beats of mistuned consonances *37* / **2.7** Fundamental tracking *40* / **2.8** Auditory coding in the peripheral nervous system *44* / **2.9** Periodicity pitch and the role of the central nervous system *50*

Sound Waves, Acoustical Energy, and the Perception of Loudness 61

3.1 Elastic waves, force, energy, and power *61* / **3.2** Propagation speed, wavelength, and acoustical power *65* / **3.3** Superposition of waves; standing waves *74* / **3.4** Intensity, sound intensity level, and loudness *78* / **3.5** The loudness perception mechanism and related processes *89*

Generation of Musical Sounds, Complex Tones, and the Perception of Tone Quality   93

4.1 Standing waves in a string *93* / **4.2** Generation of complex standing vibrations in string instruments *98* / **4.3** Sound vibration spectra and resonance *106* / **4.4** Standing longitudinal waves in an idealized air column *115* / **4.5** Generation of complex standing vibrations in wind instruments *119* / **4.6** Sound spectra of wind instrument tones *126* / **4.7** Trapping and absorption of sound waves in a closed environment *128* / **4.8** Perception of pitch and timbre of musical tones *133* / **4.9** Identification of musical sounds *138*

Superposition and Successions of Complex Tones and the Perception of Music   143

5.1 Superposition of complex tones *143* / **5.2** The sensation of musical consonance and dissonance *146* / **5.3** Building musical scales *153* / **5.4** The standard scale and the standard of pitch *158* / **5.5** Why are there musical scales and why do we experience musical sensations? *161* / **5.6** Specialization of speech and music processing in the cerebral hemispheres *165*

Appendix I   **Some Quantitative Aspects of the Bowing Mechanism**   171

Appendix II   **Some Quantitative Aspects of Recent Central Pitch Processor Models**   175

Appendix III   **Some Remarks on Teaching Physics and Psychophysics of Music**   184

References   189

Index   195

Music, Physics, and Psychophysics: An Interdisciplinary Approach

1.1
The Intervening Physical Systems

Imagine yourself in a concert hall listening to a soloist performing. Let us identify the systems that are relevant to the "music" you hear. First, obviously, we have the player and the *instrument* that "makes" the music. Second, we have the *air* in the hall that transmits the sound into all directions. Third, there is you, the *listener*. In other words, we have the chain of systems: instrument→air→listener. What links them while music is being played? A certain type and form of vibrations called *sound* which propagate from one point to another in the form of *waves* and to which our ear is sensitive. (There are many other types and forms of vibrations that we cannot detect at all, or that we may detect, but with other senses such as touch or vision.)

The physicist uses more general terms to describe the three systems listed above. He calls them: *source→medium →receptor*. This chain of systems is common to the study of many other physical processes: light, radioactivity, electricity, gravity, cosmic rays, etc. The source *emits*, the medium *transmits*, the receptor *detects*, registers, or, in general, is affected in some specific way. What is emitted, transmitted, and detected is energy—in one of its multiple forms, depending on the particular case envisaged. In the case of sound waves, it is *elastic* energy, because it involves oscillations of pressure, that is, rapidly alternating compressions and expansions of air.

Let us have a second, closer, look at the systems involved. At the source, i.e., the musical instrument, we identify several distinct components: (1) The *primary excitation mechanism* that must be activated by the player,[1] such as

[1] To make the description complete we should add the *player* and his multiple "components": the motor cortex of his brain that issues the commands to his muscles, the parts of his body with which he activates the musical instrument or his vocal tract, the feedback from ears and

the bowing or the plucking action on a violin string, the oscillating reed in a clarinet, the player's lips in a brass instrument, or the airstream blown against a wedge in the flute. This excitation mechanism acts as the primary energy source. (2) The fundamental *vibrating element* that, when excited by the primary mechanism, is capable of sustaining certain well-defined vibration modes of prefixed frequencies, such as the strings of a violin or the air column in the bore of a wind instrument or organ pipe. This vibrating element actually determines the musical pitch of the tone and, as a fortunate bonus, provides the upper harmonics needed to impart a certain characteristic quality to the tone. In addition, it serves as a vibrational energy storage. In wind instruments it partly controls the primary excitation mechanism through feedback coupling (strong in woodwinds, weak in brasses). (3) Many instruments have an additional *resonator* (sound board of a piano, body of a string instrument) whose function is to convert more efficiently the oscillations of the primary vibrating element (string) into sound vibrations of the surrounding air.

In the medium, too, we must make a distinction: we have the *medium proper* that transmits the sound and the *boundaries*, i.e., the walls, the ceiling, the floor, the people in the audience, etc., which strongly affect the sound propagation by *reflection* and *absorption* of the sound waves and whose configuration determines the quality of room acoustics (reverberation).

Finally, in the listener we single out the following principal components: (1) the *eardrum*, which picks up the pressure oscillations of the sound wave reaching the ear and converts them into mechanical vibrations that are transmitted via a link of three tiny bones to: (2) The *inner ear*, or cochlea, in which the vibrations are sorted out according to frequency ranges, picked up by receptor cells, and converted into nerve impulses. (3) The *auditory nervous system*, which transmits the neural signals to the brain where the information is processed, displayed as an image of auditory features on a certain area of the cortex (the brain surface and underlying tissue), identified, stored in the memory, and eventually transferred to other centers of the brain. These latter stages lead to the conscious perception of musical sounds. Some neural processing of acoustical information is done right at the outset in the peripheral nervous system (adaptation, contrast intensification, and, very likely, detection of transients).

Notice that we may replace the listener by a *recording device* such as a magnetic tape recorder, a phonograph

---

muscles that aids him in controlling his performance, etc. But unfortunately, limitation of space compels us to leave the player completely out of the picture.

Table 1.1

	System	Function
Source	Excitation mechanism	Energy supply
	Vibrating element	Determination of fundamental tone characteristics
	Resonator	Conversion into air pressure oscillations (sound waves), final determination of tone characteristics
Medium	Medium proper	Sound propagation
	Boundaries	Reflection, absorption, reverberation
Receptor	Eardrum	Conversion into mechanical oscillations
	Inner ear	Primary frequency sorting, conversion into nerve impulses
	Nervous system	Processing, display, identification, storage and transfer to other brain centers

recorder, or a photoelectric record on film, and still recognize at least three of the subsystems: the mechanical detection and subsequent conversion into electrical signals in the microphone, a limited amount of deliberate or accidental processing in the electronic circuitry, and the memory storage on tape, disc, or film, respectively. The first system, i.e., the instrument, of course also may be replaced by an electronic *playing device*.

We may summarize this discussion in Table 1.1

The main aim of this book is to analyze comprehensively what happens at each stage shown in Table 1.1 and during each transition from one stage to the next, when music is being played.

1.2
Characteristic
Attributes of
Musical Sounds

Subjects from all cultures agree that there are three primary sensations associated with a given single musical sound: *pitch, loudness,* and *timbre*.[2] We shall not attempt to define these subjective attributes or psychological magnitudes, or get involved at this stage in the discussion of whether they are measurable; we shall just note that pitch is fre-

[2] The sometimes quoted sensations of *volume* and *density* (or brightness) are composite concepts that can be "resolved" into a combination of pitch and loudness effects (lowering of pitch with simultaneous increase of loudness leads to a sensation of increased volume; rising pitch with simultaneous increase of loudness leads to increased density or brightness).

quently described as the sensation of "altitude" or "height," and loudness that of "strength" or "intensity" of a tone. Timbre, or quality, is what enables us to distinguish among sounds from different kinds of instruments even if their pitch and loudness are the same. The unambiguous association of these three qualities to a given sound is what differentiates a musical "tone" from "noise": although we can definitely assign loudness to a given noise, it is far more difficult to identify a unique pitch or quality.

The assignment of pitch, loudness, and timbre to a musical sound is the result of processing operations in the ear and in the brain. It is subjective and inaccessible to direct physical measurement (see Section 1.4). However, each one of these primary sensations can be associated in principle to a well-defined physical quantity of the original stimulus, i.e., the sound wave, that can be measured and expressed numerically by physical methods. Indeed, the sensation of pitch is primarily associated to the *fundamental frequency* (repetition rate of the vibration pattern, described by the number of oscillations per second), loudness to *intensity* (energy flow or pressure oscillation amplitude of the sound wave reaching the ear), and timbre to the *spectrum*, or proportion with which other, higher, frequencies called "upper harmonics" appear mixed with each other and accompany the fundamental frequency.

This, however, is a far too simplistic picture. First, the pitch sensation caused by a pure tone of fixed frequency may change slightly if we change the intensity; conversely the loudness of a tone of constant intensity will appear to vary if we change the frequency. Second, the sensation of loudness of a superposition of several tones of different pitch each is not anymore related in a simple way to the total sound energy flow; for a succession of tones of very short duration on the other hand, it depends on how long each tone actually lasts. Third, refined timbre perception as required for musical instrument recognition is a process that utilizes much more information than just the spectrum of a tone; the transient attack and decay characteristics are equally important, as one may easily verify by trying to recognize musical instruments while listening to a magnetic tape played backward. Moreover, the tones of a given instrument may have spectral characteristics that change appreciably throughout the compass of the instrument, and the spectral composition of a given tone may change considerably from point to point in a music hall; yet they are recognized without hesitation as pertaining to the same instrument. Conversely, a highly trained musician may have greatest difficulty in matching the exact pitch of a single electronically generated tone deprived of upper harmonics, fed to his ears through headphones, because his central nervous system is lacking some key additional information that normally comes with the "real" sounds to which he is exposed.

Another relevant physical characteristic of a tone is the spatial direction from which the corresponding sound wave is arriving. What really matters here is the minute time difference between the acoustical signals detected at each ear, which depends on the direction of incidence. This time difference is measured and coded by the nervous system to yield the sensation of *tone directionality* (stereophony or lateralization).

When two or more tones are sounded simultaneously, our brain is capable of singling them out individually, within certain limitations. New, less well-defined but nevertheless musically essential subjective sensations appear in connection with two or more superposed tones, collectively leading to the concept of *harmony*. Among them are the "static" sensations of *consonance* and *dissonance* describing the either "pleasing" or "irritating" character of certain superpositions of tones, the "dynamic" sensation of the *urge to resolve* a given dissonant interval or chord, the peculiar effect of *beats*, and the different character of *major* and *minor* chords. Whereas the correlation of pitch, loudness, and, to some extent, timbre with certain physical characteristics of single tones is "universal"—i.e., independent of the cultural conditioning of a given individual—this is no longer the case with the above-mentioned subjective attributes of tone superpositions.

1.3
The Time Element
in Music

A steady sound, with constant frequency, intensity, and spectrum is annoying. Moreover, after a while our conscient present wouldn't register it anymore. Only when that sound is turned off do we suddenly realize again that it had been there. Music is made up of tones whose physical characteristics change with time in a certain fashion. It is only this time dependence that makes a sound "musical" in the true sense. In general, we shall henceforth call a time sequence of individual tones, or superposition of tones, a *musical message*. Such a musical message may or may not be "meaningful" (sometimes called a tonal "Gestalt"), depending on whether we assign a certain value to it as the result of a series of brain operations of analysis, comparison with previously stored messages, storage in the memory, and associations. A *melody* is the most relevant example of a musical message. Some attributes of "meaningful" musical messages are key elements in music: *tonality* (domination of a single tone in the sequence), the *sense of return* to the tonic, *modulation*, and *rhythm*. A fundamental characteristic of a melody is that the succession of tones proceeds in discrete, finite steps of pitch in practically all musical cultures. Out of the infinite number of available frequencies, our auditory system prefers to single out discrete values corresponding to the notes of a *musical scale*, even though we are able to detect frequency changes that are much smaller than the natural step of any musical scale. The neural mechanism that analyzes a musical message pays attention only to the *transitions* of pitch; "absolute" pitch processing is lost at an early age in most individuals.

Let us examine the time element in music more closely. There are three distinct times scales on which time variations of psychoacoustical relevance occur. First, we have the "microscopic" time scale, in which the actual vibrations of a sound wave occur, covering a range from about 0.00007 to 0.05 seconds. Then there is an "intermediate" range centered at about one-tenth of a second, in which some transient changes such as tone attack and decay occur, representing the time variations of the microscopic features. Finally, we have the "macroscopic" time scale, ranging from about 0.1 second upward, corresponding to common musical tone durations, successions, and rhythm. It is important to note that each typical time scale has a particular "processing center" with a specific function, in the auditory system. (1) The microscopic vibrations are detected and coded in the *inner ear* and mainly lead to the primary tone sensations (pitch, loudness, and timbre). (2) The intermediate or transient variations seem to affect mainly processing mechanisms in the *neural pathway* from the ear to the auditory area in the brain and provide additional cues for quality perception, tone identification, and discrimination. (3) The macroscopic time changes are processed at the highest neural level—the *cerebral cortex*, the folded structure and underlying tissue forming the major part of the brain; they determine the actual musical message and its attributes.

The higher we move up through these processing stages in the auditory pathway, the more difficult it becomes to define and identify the psychological attributes to which this processing leads and the more everything appears to be influenced by learning and cultural conditioning, as well as by the momentaneous behavioral state of the individual. This gradually increasing complexity was probably dictated by the ever-increasing demands on the auditory system, first as a distance receptor and later as a communications device, in the course of phylogenetic evolution.

The "intermediate" and "macroscopic" time changes and their psychological effects have been very much neglected in experimental psychoacoustical research. For more than 100 years musicologists have bitterly complained that physics of music and psychoacoustics have been restricted mainly to the study of steady, constant tones or tone complexes, whereas the essence of music is a time sequence thereof. Their complaints are well founded, but the reasons for such a restriction are well founded, too. As explained above, the processing of tone sequences occurs at the highest level of the central nervous system—involving an as yet little explored chain of mechanisms. In this book, whenever possible, we shall try to remedy this situation and go as far as we can to dispel some of this well-placed criticism.

1.4 Physics and Psychophysics

We may describe the principal objective of physics in the following way: to provide methods by means of which one can predict quantitatively the evolution of a given physical system, based on the conditions in which the system is found initially.[3] For instance, given an automobile of a certain mass and specifying the braking forces, physics allows us to predict how long it will take to bring the car to a halt and where it will come to a stop, provided we specify the position and the speed at the initial instant of time. Given the mass, the length, and the tension of a violin string, physics predicts the possible frequencies with which the string would vibrate if plucked or bowed in a certain manner. Given the shape and dimensions of an organ pipe and the composition and the temperature of the gas inside (air), physics predicts the fundamental frequency of the sound emitted when it is blown.

In practice, "to predict" means to provide a mathematical apparatus, a series of formulas or "recipes" which, based on certain *physical laws* that govern the system under analysis, establish mathematical relationships between the values of the physical magnitudes that characterize the system at any given instant of time (position and speed in the case of the car; frequency and amplitude of oscillation in the other two examples). These relations are then used to find out how the values change as time progresses.

In order to establish the physical laws that govern a given system, we must first observe the system and make quantitative measurements of the physical magnitudes to find out their causal interrelationships experimentally. A physical law expresses a certain relationship that is common to many different physical systems and independent of particular circumstances. For instance, the law governing gravitation is valid here on Earth, on the moon, in the solar system, and elsewhere in the universe. Newton's law of motion applies to all bodies, irrespective of their chemical composition, color, temperature, speed, or position.

---

[3] The objective of physics is sometimes quoted as "the achievement of a quantitative explanation of the universe." This, however, is at best a philosophical jargon that has little to do with the actual (far more precise, realistic, and modest) scope of physics.

Most of the actual systems studied in physics—even the "simple" and "familiar" examples given above—are so complex, that accurate and detailed predictions are impossible. Thus, we must make approximations and devise simplified *models* that represent a given system only by its main features. Many times it is necessary to break up the system under study into a series of more elementary subsystems, physically interacting with each other, each one governed by a well-defined set of physical laws.

The physics of daily life's world, or *classical physics*, assumes that both measurements and predictions should always be "exact" and "unique," the only limitations and errors being due to the imperfection of our measuring or observing devices. In the atomic and subatomic domain, however, this deterministic view does not apply anymore. Nature is such that measurements and predictions of an atomic system can *never* be expected to be exact or unique in the ordinary sense, no matter how much we try to improve our techniques: measurements will always be of limited accuracy and only *probabilities*, that is, likelihood, can be predicted for the values of physical magnitudes in the atomic domain. In other words, it is impossible to predict, say, *when* a given radioactive nucleus will decay, or exactly *where* a given electron will be found at a given time during its journey from the cathode to the TV screen— only probabilities can be specified. An entirely new physics had to be constructed in the early 1920's, fit to describe atomic and subatomic systems—the so-called *quantum physics*.

The reader may wonder at this stage why we are talking about quantum physics, when it seems to be totally irrelevant to the study of sound and music. However, *psychophysics* operates in a fashion that in some aspects is surprisingly similar to quantum physics. Broadly speaking, as physics in general, psychophysics tries to make predictions on the response of a specific system subjected to given initial conditions. The system under consideration is the *brain and associated peripheral nervous and endocrine systems*, the conditions are determined by the physical *sensorial input stimuli*, and the evolution is manifested by the individual physiological reactions or by the whole complex *behavior* of the organism commanded by that brain. In man, who is able to report verbally the results of an introspective observation of his own state of mind, the evolution of the psychophysical system is usually described by the *psychological sensations* and feelings evoked by the sensorial stimuli (this is sometimes called sensory psychophysics in distinction to the above motor psychophysics). Like physics, psychophysics requires that the causal relationship between physical stimulus input and behavioral or psychological output be established through experimentation and measurement. Like physics, psychophysics must make simplifying assumptions and construct *models*—mainly for nervous and neuropsychological systems operations—in order to be able to venture into the business of prediction-making.

Unlike classical physics, but strinkingly similar to quantum physics, psychophysical predictions can never be expected to be exact or unique—only probability values can be established. Unlike classical physics, but strikingly similar to quantum physics, most measurements in psychophysics will substantially perturb the system under observation, and nothing can be done to eliminate said perturbation completely. As a consequence, the result of a measurement does not reflect the state of "the system per se," but, rather, the more complex state of "the system under observation." Also as a result, psychophysics requires experimentation with many different equivalent (but never identical) systems (subjects), and a *statistical interpretation* of the results.[4]

Quite obviously, there are some limits to these analogies. In physics, the process or "recipe" of the measurement which defines a given physical magnitude, such as the length, weight, or velocity of an object, can be formulated in a rigorous, unambiguous way. As long as we deal with physiological output, such as neural impulse rate, amplitude of evoked goosepimples or increase in heartbeat, measurements can be formulated in a rigorous, unambiguous way, too. But in sensory psychophysics, how do we define and measure the subjective sensations of pitch, loudness, or—to make it even more tricky—the magnitude that represents the urge to bring a given melody to its tonic completion? Or how would we arrange measurements on "internal hearing," i.e., the action of provoking musical tone images be volition, without external stimuli? Could this be done by interrogation alone, or must we resort to "direct" measurements via implantation of microelectrodes into brain cells?

It is now believed that a given sensation is related to neural activity, evoked by sensory input signals, and "displayed" on the cortical area to which the stimulated sense organ is wired (primary auditory receiving area, visual cortex, etc.). Whenever the "read-out" mechanism that represents our conscient present, and whose function is to monitor and synthesize the image of environmental features continuously being displayed on the sensory cortex areas,

---

[4] We must emphasize that these are only *analogies*. Quantum physics *as such* does not play an explicit role in the nervous system, whose operation is based on eminently classical processes (although some scientists believe that individual quantum-mechanical interactions may play an explicit role in the mechanism of transmission of neural signals from one cell to another).

directs its attention to the particular area in question, the encountered activity gives rise to what we report as a "sensation." The general location and the spatial and temporal distribution of this neural activity determine the class and the subjective "intensity" of the associated sensation. Many sensations can indeed be *classified* into more or less well-defined types (called sensorial qualities, if they are caused by the same sense organ)—the fact that people do report to each other on pitch, loudness, tone quality, consonance, etc., without much mutual misunderstanding with regard to the meaning of these concepts, is an example. Furthermore, two sensations belonging to the same type, experienced one following the other, can in general be *ordered* by the experiencing subject as to wether the specific attribute of one is felt to be "greater" (or "higher," "stronger," "brighter," "more pronounced," etc.), "equal," or "less" than the other. For instance, when presented with two tones in a succession, the subject can judge whether the second tone was of higher, equal, or lower pitch than the first one. Another example of ordering is the following: presented with the choice of three complex tones of the same pitch and loudness, he may order them in pairs by judging which two tones have the most similar timbre and which the most dissimilar one. One of the fundamental tasks of psychophysics is the determination, for each type of sensation, of the minimum detectable value (or threshold value) of the physical magnitude responsible for the stimulus, and the minimum detectable change (the "just noticeable difference" or *jnd*).

The ability, possessed by all individuals, to classify and order subjective sensations gives subjective sensations a status quite equivalent to that of a physical magnitude and justifies the introduction of the term *psychophysical magnitude*. What we must *not* expect a priori is that individuals can judge without previous training whether a sensation is "twice" or "half," or any other *numerical* factor, that of a reference unit sensation. There are situations, however, in which it is possible to *learn* to make quantitative estimates of psychophysical magnitudes on a statistical basis, and, in some instances, the brain may become very good at it. The visual sense is an example. After sufficient experience, the estimation of the size of objects can become highly accurate, provided enough information about the given object is available; judgments such as "twice as long" or "half as tall," are made without hesitation. It is quite clear from this example that a "unit" and the corresponding psychophysical process of comparison have been built into the brain only *through experience and learning*, in multiple contacts with the original physical magnitudes. The same can be achieved with other psychophysical sensations such as loudness: it is necessary to acquire through learning the ability of comparison and *quantitative* judgment. The fact that musicians all over the world use a common loudness notation is a self-evident example.

And here we come to the perhaps most crucial differences between physics and psychophysics: (1) repeated measurements of the same kind may *condition the response* of the psychophysical system under observation: the brain has the ability of learning, gradually changing the course of response probability to a given input stimulus, as the number of similar exposures increases. (2) The *free will* of the subject under study and the consequences thereof, mental or physical, may interfere in a highly unpredictable way with the measurements. As a consequence of the first point, a statistical psychophysical study with one single individual exposed to repeated "measurements" will by no means be identical to a statistical study involving one single measurement performed on each one of many different individuals. This difference is due not only to differences among individuals, but also to the conditioning that takes place in the case of repeated exposures. The ultracomplex feedback loops in the nervous system make psychoacoustical measurements particularly tricky to set up and to interpret.

1.5
What Is Music?

The previous discussion may have irritated some readers. Music, they will say, is "pure aesthetics," a manifestation of the innate and sublime human comprehension of the beautiful rather than the mere effect of certain sound wave stimuli on a complex network of billions of nerve cells. However, ultimately *even aesthetic feelings must be somehow related to neural information processing*. The so-characteristic blend of regular, ordered patterns alternated with surprise and uncertainty, common to all sensorial input judged as "aesthetic," may be a manifestation of man's curious, yet fundamental desire to exercise his superredundant neural network with biologically nonessential information-processing operations of changing or alternating complexity. Indeed, artistic creativity is perhaps the most human of all intellectual qualities. Whereas it might be argued that intelligence and the capacity to communicate are only higher in *degree* in humans than in animals, artistic creativity and appreciation are absolutely unique to human beings.[5]

Music seems to be a quite natural by-product of the evolution of speech and language. In this evolution, which undoubtedly *was* an essential factor for the development of the human race, a neural network emerged, capable of executing the ultracomplex sound processing, identification, storage, and retrieval operations necessary for phonetic recognition, voice identification, and comprehension of

[5] Obviously, we do not subscribe to the opinion that plants, cows, and chickens, when exposed to this or that kind of music, raise their productivity because of artistic appreciation!

words and sentences. Language endowed humans with a mechanism that increased the capacity of their memory (and associated storage, retrieval, and communications operations) billions of times, by allowing the reduction of ultracomplex *images* of environmental scenes, objects, and their causal interrelations to short *symbolic* representations. In the course of this evolution, it so happened that a most remarkable division of tasks between the two cerebral hemispheres developed (Section 5.6). The left hemisphere (in about 97% of all persons) mainly executes short-term temporal operations such as required for verbal intelligibility and other short-term *sequencing operations*, such as thinking. The right hemisphere cooperates with the performance of *spatial integrations* and long-term time representations. Examples of these *holistic* operations of the right hemisphere are pictoral imaging—and *musical perception*. Indeed, as we shall see throughout this book, musical perception does involve the analysis of spatial excitation patterns along the auditory receptor organ, caused by musical tones and tone superpositions, as well as the analysis of long-term time patterns of melodic lines.

Why do we respond emotionally to complex musical messages which seem to contain no information of any real survival value? The fact that most of us do—often without any special training—indicates that *the human brain is instinctively motivated to entertain itself with sound-processing operations even if such activity is not required by momentaneous environmental circumstances.* This motivation may well be the result of an inborn drive to train at an early age in the highly sophisticated auditory analysis operations expected for speech perception—not unlike an animal's play being the manifestation of an inborn motivation to develop or improve skilled movements required for preying and self-defense.

Because the perception of music is ultimately based on acoustical information processing, the major or minor degree of complexity of a tone-message identification, the degree of success of prediction-making operations that are carried out by the brain to expedite this identification process,[6] and the type of associations evoked by comparison with stored information on previous experiences may be the ultimate combined "cause" of the musical sensations evoked by a given musical message. If this is true, it would be quite obvious that both innate neural mechanisms (primary processing operations) *and* cultural conditioning (stored messages and learned processing operations) must control our behavioral and aesthetic response to music.

----

[6] We should point out that, quite generally, the development of a high-speed prediction-making capability must have been a leading motivation for the evolution of the central nervous system, designed to endow the higher species with predictor–corrector mechanisms that were necessary to enhance their chance of survival in an ecological environment of ever-increasing complexity.

2 Sound Vibrations, Pure Tones, and the Perception of Pitch

We hear a "sound" when the eardrum is set into a characteristic type of periodic motion called *vibration*. This vibration is caused by small pressure oscillations of the air in the auditory canal associated to an incoming sound wave. In this chapter we shall first discuss the fundamentals of vibratory motion in general and then focus on the effects of eardrum vibrations on our sensation of hearing. We shall not worry at this stage about *how* the eardrum is set into motion. To that effect, let us imagine that we put on headphones and listen to tones generated therein. In the lower frequency range the eardrums will follow very closely the vibrations of the headphone diaphragms. This approach of introducing the subject is somewhat unorthodox. But it will enable us to plunge straight into the study of some of the key concepts associated with sound vibration and sound perception without spending first a long time on sound waves and sound generation. From the practical point of view, this approach has one drawback: the experiments that we shall present and analyze in this chapter necessarily require electronic generation of sound rather than natural production with real musical instruments. Whenever possible, however, we shall indicate how a given experiment could be performed with real instruments.

2.1 Motion and Vibration

Motion means *change of position* of a given body with respect to some reference body. If the moving body is very small with respect to the reference body, or with respect to the dimensions of the spatial domain covered in its motion, so that its shape is practically irrelevant, the problem is reduced to the description of the motion of a *point* in space. This is why such a small body is often called a material point or a particle. On the other hand, if the body is not small, but if from the particular circumstances we know beforehand that all points of the body are confined to move along straight lines parallel to each other ("rectilinear translation"), it, too, will suffice to specify the motion of just *one* given point of the body. This is a "one-dimensional" case of motion, and the position of the given point of the body (and, hence, that of the whole body) is completely specified by just *one* number: the distance to a fixed reference point.

In this book we shall only deal with one-dimensional motions. Let us assume that our material point moves along a vertical line (Fig. 2.1). We shall designate the reference point on that line with the letter O. Any fixed point can serve as a reference point, although, for convenience, we sometimes may select a particular one (such as the equilibrium position for a given oscillatory motion). We indicate the position of a material point P by the distance y to the reference point O (Fig. 2.1). y is also called the *displacement* of P with respect to O, or the *coordinate* of P. We must use both positive and negative numbers to distinguish between the two sides with respect to O.

In physics, the *metric system* is used to measure distances. The unit of length is the *meter* (1 m = 3.28 feet); several decimal submultiples (e.g., the centimeter = 0.01 m = 0.394 inch, or the millimeter = 0.001 m) and multiples (e.g., the kilometer = 1000 m = 0.625 mile) are also frequently used.

The material point P is in motion with respect to O when its position changes with time. We shall indicate time by the letter t. It is measured with a clock—and it, too, requires that we specify a "reference" instant of time at t = 0. Motion can be represented mathematically in two ways: analytically, using so-called functional relationships, and geometrically, using a graphic representation. We shall use only the geometric method. To represent a one-dimensional

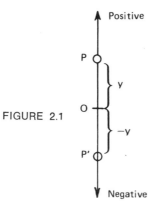

FIGURE 2.1

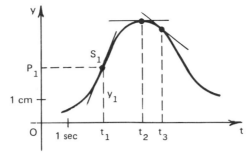

FIGURE 2.2

motion graphically, we introduce two axes perpendicular to each other, one representing the time t, the other the coordinate y (Fig. 2.2). For both we have to indicate clearly the *scale*, i.e., the unit intervals (of time and displacement, respectively). A motion can be represented by plotting for each instant of time t, the distance y at which the particle is momentarily located. Each point of the ensuing curve, such as S_1, (Fig. 2.2), tells us that $t = t_1$ the particle P is at a distance y_1 from O, i.e., at position P_1. Note very carefully that in this graph the material point does *not* move along the curve of points S. This curve is just a human-made "aid" that helps us to find the position y of the particle at any time t.

The graph shown in Fig. 2.2 also gives information on the velocity of the material point, i.e., the rate of change of its position. This is determined by the *slope* of the curve in the graph: at t_1 the particle is moving at a certain rate upward, at t_3 it is moving downward at a slower rate. At t_2 it is momentarily at rest, reversing its direction.

There is a certain class of motions in which the material point follows a pattern that is repeated time and again. This is called *periodic motion*, or *vibration*. It is the type of motion of greatest importance to physics of music. In order to have a truly periodic motion, a body not only has to come back to the same position repeatedly, it has to do so at exactly equal intervals of time and repeat exactly the same type of motion in between. The interval of time after which the pattern of motion is repeated is called the *period* (Fig. 2.3a). We denote it by the Greek letter tau (τ). During one period the motion may be very simple (Fig. 2.3a) or rather complicated (Fig. 2.3b). The elementary pattern of motion that occurs during one period and that is repeated over and over again is called a *cycle*.

There are mechanical and electronic devices that can automatically plot the graph of a periodic motion. In a *chart-recorder*, the pen reproduces in the y direction the periodic motion that is to be described, while it is writing on a strip of paper that is moving perpendicularly to the

(a)

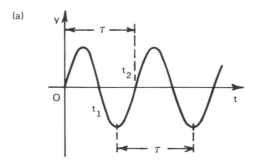

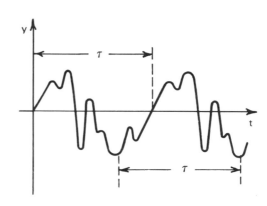

FIGURE 2.3

y axis at a constant speed. Since we know this speed, we can assign a *time* scale to the axis along the paper strip. The curve obtained is the graphic representation of the motion. This method is not practical for the registration of acoustical vibrations though. They have such short periods that it would be impossible to displace a pen fast enough to reproduce this type of vibration. An electronic device called *oscilloscope* serves the purpose. In essence, it consists of a very narrow beam of electrons (elementary particles of negative electric charge) that impinge on a TV-type screen giving a clearly visible light spot. This beam can be deflected both in the vertical and in the horizontal directions. The vertical motion is controlled by a signal proportional to the vibration which we want to display (for instance, the vibration of the diaphragm of a microphone). The horizontal motion is a continuous sweep to the right with constant speed, equivalent to the motion of the paper strip in a chart recorder, thus representing a time scale. The luminous point on the screen thus describes the graph of the motion during one sweep. If the image of the luminous point is retained long enough, it appears as a continuous curve on the screen. Since the screen is only of limited size, the horizontal motion is instantaneously reset to the

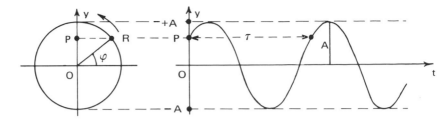

FIGURE 2.4

origin whenever the beam reaches the right edge of the screen, and the sweep starts again.

2.2 Simple Harmonic Motion

The question now arises as to which is the "*simplest*" kind of periodic motion. There are many examples in nature: the back and forth oscillation of a pendulum, the up and down motion of a spring, the oscillations of molecules, etc. Their motions have something important in common: they all can be represented as the projection of a uniform circular motion onto one diameter of the circle (Fig. 2.4).[1] When point R turns around uniformly (with period τ, i.e., once every τ seconds) the projection point P moves up and down along the y axis with what is called a *simple harmonic motion* (see graph at right side of Fig. 2.4). This is also called a *sinusoidal motion* (because y can be represented analytically by a trigonometric function called sine).

Note that a simple harmonic motion represents a vibration that is symmetric with respect to point O, which is called the equilibrium position. The maximum displacement A (either up or down) is called *amplitude*. τ is the *period* of the harmonic motion. There is one more parameter describing simple harmonic motion, that is a little more difficult to understand. Consider Fig. 2.4: at the initial instant $t = 0$, the particle (projection of the rotating point R) is located at position P. We may now envisage a second case of harmonic motion with the *same* period τ and the *same* amplitude A, but in which the particle starts from a *different position Q* (Fig. 2.5). The resulting motion obviously will be different, not in form or type, but in relative "timing." Indeed, as seen in Fig. 2.5, both particles will pass through a given position (e.g., the origin O) at different times (t_1, t_2). Conversely, both particles will in general be at different positions at one given time (e.g., P and Q at

---

[1] Note carefully that the construction at the left side of Fig. 2.4 is *auxiliary*; the only real motion is the periodic up-and-down motion of the particle P along the y axis.

FIGURE 2.5

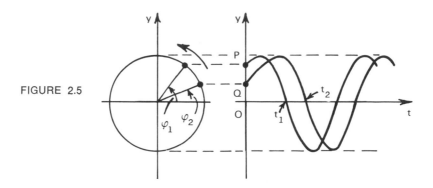

$t = 0$). If we again imagine the motion of the second particle Q as the projection of a uniform circular motion (Fig. 2.5), we realize that both cases pertain to different *angular positions* φ_1, φ_2 of the associated rotating points on the circle. The angle φ is called the *phase* of a simple harmonic motion; the difference $\varphi_1 - \varphi_2$ (Fig. 2.5), which remains constant in this example, is called the *phase difference* between the two harmonic motions.

In summary, a given "pure," or harmonic, vibration is specified by the values of three parameters: the *period* τ, the *amplitude* A, and the *phase* φ (Fig. 2.4). They all, but especially the first two, play a key role in the perception of musical sounds.

Simple harmonic motions occur practically everywhere in the universe: vibrations of the constituents of the atoms, of the atoms as a whole in a crystal, of elastic bodies, etc., can all be described in terms of simple harmonic motions. But there is another even more powerful reason for considering simple harmonic motion as the most basic periodic motion of all: it can be shown mathematically that *any kind of periodic motion, however complicated, can be described as a sum of simple harmonic vibrations*. We shall deal with this fundamental property later on in detail (Chapter 4). It is indeed of capital importance to music.

2.3 Acoustical Vibrations and Pure Tone Sensations

When the eardrum is set into periodic motion, its mechanical vibrations are converted in the inner ear into nerve impulses that are signaled to the brain and interpreted as sound, provided the period and the amplitude of the vibrations lie within certain limits. In general, the ear is an extremely sensitive device: vibration amplitudes of the eardrum as small as 10^{-7} cm can be detected, and so can vibrations with periods as short as 7×10^{-5} sec.[2]

---

[2] In this book we shall use the exponent notation: $10^{+n} = 100\ldots\ldots\ldots00$. (⌒ n zeroes ⌒) 10^{-n} is just $1/10^n$, that is, a decimal fraction given by one unit in the nth decimal position.

We now introduce a quantity that is used more frequently than the period τ, called the *frequency:*

$$f = \frac{1}{\tau} \tag{2.1}$$

Physically, f represents the number of repetitions of the vibration pattern, or cycles, in the unit of time. The reason for preferring f to τ is that the frequency increases when our sensation of "tone-height" or pitch increases. If τ is given in seconds, f is expressed in *cycles per second*. This unit is called hertz (Hz), in honor of Heinrich Hertz, the famous German physicist. Vibrations in the interval 20 Hz -15,000 Hz are sensed as "sound" by a normal person. Both the lower and particularly the upper limit depend on tone loudness and vary considerably from person to person and with age.

When a sound causes a simple harmonic motion of the eardrum with constant characteristics (frequency, amplitude, phase), we hear what is called a *pure tone*. A pure tone sounds dull, and music is not made up of single pure tones. However, as stated in the introduction to this chapter, for a better understanding of complex sounds, it is advisable to deal first with pure or simple tones only. Pure tones have to be generated with electronic oscillators; there is no musical instrument that produces them (and even for electronically generated pure tones, there is no guarantee that they will be "pure" when they actually reach our ear). In any case, since the flute is the instrument whose sound approximates that of a pure, sinusoidal tone more than any other instrument, especially in the upper register, several (but not all) of the experiments referred to in this chapter can indeed be performed at home using one or, as a matter of fact, two flutes—played by experts, though!

When we listen to a pure tone whose frequency and amplitude can be changed at will, we verify a correspondence between *pitch* and *frequency* and between *loudness* and *amplitude*. One has a fairly good idea on how the ear's primary frequency and amplitude detection mechanism works for pure sounds. In this chapter we only consider pitch. The simple harmonic oscillations of the eardrum are transmitted by a chain of three small bones (called hammer, anvil, and stirrup) to a membrane at the entrance (oval window) to the *cochlear duct*, which represents the inner ear proper (Fig. 2.6). This duct, wound up like a snail's shell (in Fig. 2.6b it is shown "stretched out"), is partitioned longitudinally into two tubes by the *basilar membrane*, about 3.5 cm long, which holds the actual sensor organ (organ of Corti) and corresponding nerve endings. Both sections of the cochlea are connected at the far end, or apex, by a small hole in the basilar membrane called

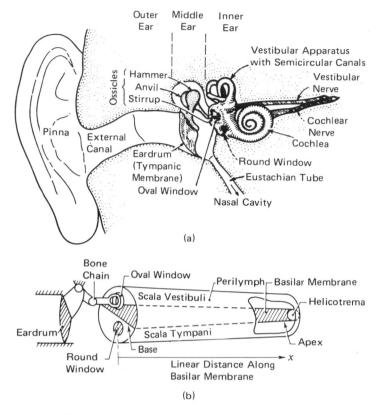

FIGURE 2.6. (a) Schematic view of the ear (Flanagan 1972,
Fig. 4) (not in scale); (b) the cochlea shown stretched
out (highly simplified).

helicotrema (Fig. 2.6b). They are filled with an incompressi-
ble fluid, the *perilymph*.[3] The lower section is sealed off
with another elastic membrane at the round window.

The vibrations transmitted by the bone chain to the oval
window membrane are converted into oscillations of the
perilymph fluid. These oscillations propagate through the
cochlear duct, and the basilar membrane is set into motion
like a waving flag. About 30,000 receptor units, called *hair
cells,* arranged in "inner" and "outer" rows along the basilar
membrane (Fig. 2.7) (Bredberg et al. 1970) pick up the motions of the latter and impart signals to the nerve cells, or
neurons, that are in contact with them (Spoendlin 1970).

[3] This is a highly oversimplified description. In reality, the cochlear
partition is highly structured, including a channel (the scala media)
filled with the endolymphatic liquid, bound by the basilar membrane
and by Reissner's membrane. Another membrane (tectorial membrane) "floats" on top of the basilar membrane and plays a key role
in the actual stimulation process.

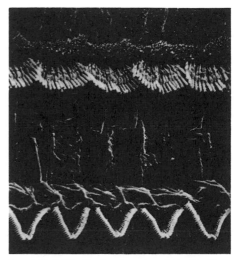

FIGURE 2.7

Scanning electron micrograph (Bredberg et al. 1970) of the inner row (*top*) and outer row (*bottom*—only one of three rows is shown) of hair cells on the basilar membrane of a guinea pig.

The remarkable fact is that for a pure tone of given frequency, the maximum basilar membrane oscillations occur only in a given, limited region of the membrane, *whose position depends on the frequency of the tone.* In other words, for each frequency there is a region of maximum sensitivity, or "resonance region," on the basilar membrane. The lower the frequency of the tone, the closer to the apex (Fig. 2.6b) lies the region of activated hair cells (where the membrane is most flexible). The higher the frequency, the closer to the entrance (oval window) it is located (where the membrane is stiffest). *The spatial position x along the basilar membrane (Fig. 2.6b) of the responding hair cells and associated neurons determines the primary sensation of pitch.* A change in frequency of the pure tone causes a shift of the position of the activated region; this shift is then interpreted as a change in pitch. We say that the primary information on tone frequency is "coded" by the sensorial organ of the basilar membrane in the form of *spatial location* of the activated neurons. Depending on which group of neural fibers is activated, the pitch will appear to us as low or high.

Figure 2.8 shows how the position x (measured from the base, Fig. 2.6b) of the region of maximum sensitivity varies with the frequency of a pure, sinusoidal tone, for an average adult person (von Békésy 1960). Several important conclusions can be drawn. First of all, note that the musically most important range of frequencies (approximately 20–4,000 Hz) covers roughly two-thirds of the extension of the basilar membrane (12–35 mm from the base). The large

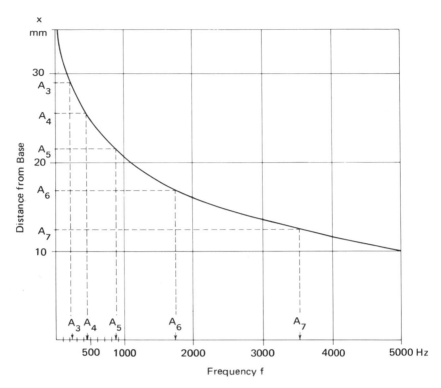

FIGURE 2.8. Position of the resonance maximum on the basilar membrane (after von Békésy 1960) for a pure tone of frequency f (linear scales).

remaining portion of the frequency scale (4,000–16,000 Hz, not shown beyond 5,000 Hz in Fig. 2.8) is squeezed into the remaining one-third. Second, notice the significant fact that whenever the frequency of a tone is doubled, i.e., the pitch jumps one octave, the corresponding resonance region is displaced by a roughly constant amount of 3.5–4 mm, no matter whether this frequency jump is from 220 to 440 Hz, from 1,760 to 3,520 Hz, or, as a matter of fact, from 5,000 to 10,000 Hz. In general, whenever the frequency f is multiplied by a given factor, the position x of the resonance region is not multiplied but simply shifted a certain amount. In other words, it is frequency *ratios*, not their differences, that determine the displacement of the resonance region along the basilar membrane. A relationship of this kind is called "logarithmic" (Section 3.4).

The above results come from physiological measurements performed on dead (but well preserved) animals (von Békésy 1960). Today such measurements can be performed on living cochleas through the Mössbauer effect. A minute mass of radioactive substance (cobalt 57) is "implanted" on the basilar membrane. The tiny displacements of the membrane can then be detected indirectly by measuring the frequency shift (Doppler effect) of gamma rays emitted by the substance (e.g., Rhode and Robles 1974).

We now consider the psychophysical magnitude pitch, associated to a pure tone of frequency *f*. In Section 1.4 we mentioned that a psychophysical magnitude cannot be measured in the same quantitative manner as a physical magnitude, such as frequency. Only a certain *order* can be established by the experiencing individual between two sensations of the same kind presented in immediate succession. Quantitative estimates are possible only after the brain somehow has been trained to perform the necessary operations (e.g., a child learning to estimate the size of the objects he sees)—and the results would have to be interpreted in a statistical way.

Let us consider an individual's ability to establish a relative order of pitch when two pure tones (of the same intensity) are presented one after the other. There is a natural limit: when the difference in frequency between the two tones is too small, below a certain value, both tones are judged as having the same pitch. This is true with the order judgments for all psychophysical magnitudes: whenever the variation of an original physical stimulus lies within a certain "difference limen" or *just noticeable difference (jnd)*, the associated sensation is judged as remaining "the same"; as soon as the variation exceeds the *jnd,* a change in sensation is detected. Notice that the *jnd* relates to a *physical* magnitude (the stimulus), is measurable in the ordinary sense, and is expressed by a number.

The degree of sensitivity of the primary pitch perception mechanism to frequency changes, or *frequency resolution* capability, depends on the frequency, intensity, and duration of the tone in question—and on the suddenness of the frequency change. It varies greatly from person to person, is a function of musical training, and unfortunately, *depends considerably on the method of measurement employed.* Figure 2.9 shows the average *jnd* in frequency for pure tones of constant intensity (80 decibels, Section 3.4), whose frequency was slowly and continuously modulated up and down (Zwicker, Flottorp, and Stevens 1957). This graph shows for instance, that for a tone of 2,000 Hz a change of 10 Hz—ie., of only 0.5%—already can be detected. This is a very small fraction of a semitone! *Sudden* changes in frequency are detected with a considerably lower *jnd*—up to 30 times smaller than the values shown in Fig. 2.9 (Rakowski 1971). Frequency resolution becomes worse at low frequencies (e.g., 3% at 100 Hz in Fig. 2.9). It also decreases with decreasing tone duration once the latter falls below about one-tenth of a second. In con-

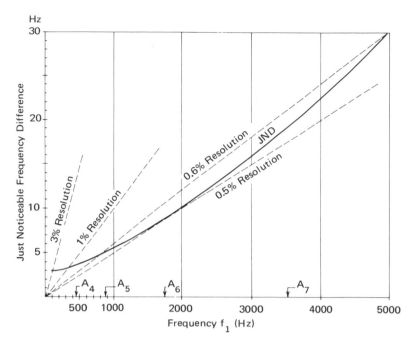

FIGURE 2.9. Just noticeable difference (*jnd*) in frequency for a pure tone of frequency f_1 (linear scales), as determined with a slowly frequency-modulated signal (after Zwicker, Flottorp, and Stevens 1957).

trast, frequency resolution is roughly independent of amplitude (loudness).

Since the inception of psychophysics, psychologists have been tempted to consider the minimum perceptible change in sensation caused by a *jnd* of stimulus as the "natural" unit with which to "measure" the corresponding psychophysical magnitude. The minimum perceptible change in pitch has been used to construct a "subjective scale of pitch" (Stevens, Volkmann, and Newmann 1937). However, as we shall see later, because the *octave* plays such an overriding role as a "natural" pitch interval, and because all musical scales developed in total independence of the attempts to establish a subjective scale of pitch, the latter has not found a direct practical application in music (see, however, Section 5.4).

In summary, according to the description given in this section, the primary function of the inner ear (cochlea) is to convert a vibration pattern in time (that of the eardrum) into a vibration pattern in space (along the basilar membrane), and this, in turn, into a spatial pattern of neural activity. The theoretical description of this mechanism is called the *Place Theory of Hearing*. We shall refer to it quite frequently and shall see that it is a good theory, but not a complete one.

2.4 Superposition of Pure Tones: First Order Beats and the Critical Band

We said before that single pure tones sound very dull. Things become a little livelier as soon as we superpose two pure tones by sounding them together. In this section we shall analyze the fundamental characteristics of the superposition of two pure tones. We will meet some very fundamental concepts of physics of music and psychoacoustics.

There are two kinds of superposition effects, depending on where they are processed in the listener's auditory system. If the processing is *mechanical,* occurring in the cochlear fluid and along the basilar membrane, we call them "first order superposition effects," mainly because they are clearly distinguishable and of fairly easy access to psychoacoustical experimentation. "Second order" superposition effects are the result of *neural* processing and are more difficult to detect, describe, and measure unambiguously. In this section we shall focus on first order effects only.

Let us first discuss the physical meaning of "superposition of sound." The eardrum moves in and out commanded by the pressure variations of the air in the auditory canal. If it is ordered to oscillate with a pure harmonic motion of given frequency and amplitude, we hear a pure tone of certain pitch and loudness. If now *two* pure tones of different characteristics are sounded together (e.g., by listening to two independent sources at the same time), the eardrum reacts as if it were executing at the same time two independent commands simultaneously, one given by each pure tone. The resulting motion is the sum of the individual motions that would occur if each pure sound were present alone, in the absence of the other. Not only the eardrum behaves this way, but also the medium and all other vibrating components (this, however, is not entirely true if the amplitudes are very large). This effect is called a *linear superposition* of two vibrations. Linear superposition of two vibrations is a technical term that means "peaceful coexistence": one component vibration does not perturb the affairs of the other, and the resulting superposition simply follows the dictations from each component simultaneously. In a *non*linear superposition, the dictation from one component would depend on whatever the other one has to say, and vice versa.

We start our discussion with the analysis of the superposition of two simple harmonic motions with *equal frequency* and *equal phase* (zero phase difference, Section 2.2). It can be shown graphically (Fig. 2.10) and also analytically that in this case we again obtain a simple harmonic motion of the same frequency, the same phase, but with an amplitude which is the sum of the amplitudes of the

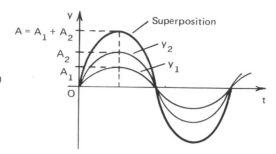

FIGURE 2.10

two component vibrations. If the two component oscillations of given frequency have *different phase,* their superposition still turns out to be a simple harmonic motion of
the same frequency, but the amplitude will *not* be given
anymore by the sum of the component amplitudes. In particular, if the amplitudes of the component vibrations are
equal and their phase difference φ is 180°, both oscillations
will annihilate each other and no sound at all will be heard.
This is called *destructive interference* and plays a role in
room acoustics. In summary, when two pure tones of equal
frequency reach the eardrum, we perceive only *one* tone of
given pitch (corresponding to the frequency of the component tones) and loudness (controlled by the amplitudes of
the superposed tones and their phase difference).

We now consider the superposition of two simple tones
of the same *amplitude* but with *slightly different frequen-
cies,* f_1 and $f_2 = f_1 + \Delta f$. The frequency difference Δf has a
small value. Let us assume that it is positive; the tone corresponding to f_2 is thus slightly sharper than that of f_1. The
vibration pattern of the eardrum will be given by the sum
of the patterns of each component tone (Fig. 2.11). The result of the superposition (heavy curve) is an oscillation of
period and frequency *intermediate* between f_1 and f_2, and
of slowly modulated amplitude. Note in this figure the
slowly changing phase difference between the component
tones y_1 and y_2: they start in phase (0° phase difference, as
in Fig. 2.10) at the instant $t = 0$, then y_2 starts leading in
phase (ahead of y_1) until both are completely out of phase
(180° phase difference) at the instant corresponding to C.
The phase difference keeps increasing until it reaches
$360° = 0°$ at instant τ_B. This continuous, slow phase shift is
responsible for the changing amplitude of the resultant
oscillation: the broken curves in Fig. 2.11 represent the
amplitude envelope of the resulting vibration (see also
Fig. 2.17A).

What is the resulting tone sensation in this case? First of
all, note very carefully that the eardrum will follow the oscillation as prescribed by the *heavy* curve in Fig. 2.11. The

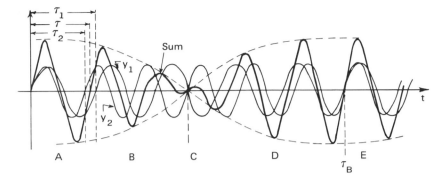

FIGURE 2.11

eardrum "does not know" and "does not care" about the
fact that this pattern really is the result of the sum of two
others. It has just *one* vibration pattern, of varying amplitude. A most remarkable thing happens in the cochlear
fluid: this rather complicated, but single, vibration pattern
at the oval window gives rise to *two* resonance regions of
the basilar membrane. *If the frequency difference Δf between the two component tones is large enough,* the corresponding resonance regions are sufficiently separated
from each other; each one will oscillate with a frequency
corresponding to the component tone (light curves in Fig.
2.11), and *we hear two separate tones of constant loudness,* with pitches corresponding to each one of the original
tones. This property of the cochlea to disentangle a complex vibration pattern caused by a tone superposition into
the original pure tone components is called *frequency discrimination.* It is a mechanical process, controlled by the
hydrodynamic and elastic properties of the inner ear constituents. On the other hand, *if the frequency difference Δf
is smaller than a certain amount,* the resonance regions
overlap, and *we hear only one tone of intermediate pitch
with modulated or "beating" loudness.* In this case, the
overlapping resonance region of the basilar membrane follows a vibration pattern essentially identical to that of the
eardrum (heavy curve in Fig. 2.11). The amplitude modulation of the vibration pattern (envelope shown in Fig. 2.11)
causes the perceived loudness modulation. We call this
phenomenon "first order beats." These are the ordinary
beats, well known to every musician.

The frequency of the resulting vibration pattern of two
tones of very similar frequencies f_1 and f_2 is equal to the
average value:

$$f = \frac{f_1 + f_2}{2} = f_1 + \frac{\Delta f}{2} \tag{2.2}$$

The time interval τ_B (Fig. 2.11), after which the resulting amplitude attains the initial value, is called the beat period. The beat frequency $f_B = 1/\tau_B$ (number of amplitude changes per second) turns out to be given by the difference:

$$f_B = f_2 - f_1 = \Delta f \tag{2.3}$$

It makes no difference whether f_2 is greater than f_1, or vice versa. Beats will be heard in either case, and their frequency will always be given by the frequency difference of the component tones (relation 2.3 really must be taken in *absolute value*, i.e., positive only). The closer together the frequencies f_1 and f_2 are, the "slower" the beats will result. If f_2 becomes equal to f_1, the beats disappear completely: both component tones sound in *unison*.

Let us summarize the tone sensations evoked by the superposition of two pure tones of equal amplitude and of frequencies f_1 and $f_2 = f_1 + \Delta f$, respectively. To that effect, let us assume that we hold f_1 steady and increase f_2 slowly from f_1 (unison, $\Delta f = 0$) to higher values. (Nothing would change qualitatively in what follows, if we were to decrease f_2.) At unison, we hear one single tone of pitch corresponding to f_1 and a loudness that will depend on the particular phase difference between the two simple tones. When we slightly increase the frequency f_2, we continue hearing *one* single tone, but of slightly *higher* pitch, corresponding to the average frequency $f = f_1 + \Delta f/2$ (2.2).[4] The loudness of this tone will be beating with a frequency Δf (2.3). These beats increase in frequency as f_2 moves away from f_1 (Δf increases). As long as Δf is less than about 10 Hz, these beats are perceived very clearly. When the frequency difference Δf exceeds, say, 15 Hz, the beat sensation disappears, giving way to a quite characteristic *roughness* or unpleasantness of the resulting tone sensation. When Δf surpasses the so-called *limit of frequency discrimination* Δf_D (not to be confused with the limen of frequency resolution, or *jnd* in frequency, of Fig. 2.9), we suddenly distinguish *two* separate tones, of pitches corresponding to f_1 and f_2. At that moment, both resonance regions on the basilar membrane have separated from each other sufficiently to give two distinct pitch signals. However, at that limit the sensation of roughness still persists, especially in the low pitch range. Only after surpassing a yet larger frequency difference Δf_{CB} called the *critical band*, the roughness sensation disappears, and both pure tones sound "smooth" and "pleasing." This transition

---

[4] How do we verify that, indeed, the pitch sensation of the resulting tone *does* correspond to a tone of frequency $f_1 + \Delta f/2$? This is accomplished with *pitch matching* experiments: the subject is presented alternatively with a *reference* tone of controllable pitch and is requested to "zero in" the frequency of the latter until he senses "equal pitch" with respect to the original tone.

from "roughness" to "smoothness" is in reality more gradual; the critical band as defined here only represents the
approximate frequency separation at which this transition
takes place.

All these results are easily verified using two electronic
"sine-wave generators" of variable frequency whose output is combined and fed monaurally into each ear with
headphones. But they can also be verified, at least qualitatively, with two flutes played simultaneously in the upper
register by expert players. While one flautist maintains a
fixed tone (holding the pitch very steady), the other plays
the same *written* note out of tune (pulling out or pushing
in step by step the mouthpiece). Beats, roughness, and tone
discrimination can be explored reasonably well.

· Figure 2.12 is an attempt (not in scale) to depict the
above results comprehensively. The heavy lines represent
the frequencies of tones (or beats) that are *actually* heard.
Tone f_1 is that of fixed frequency, f_2 corresponds to the
tone whose frequency is gradually changed (increased or
decreased). The "fused" tone corresponds to the single
tone sensation (of intermediate frequency) that is perceived
as long as f_2 lies within the limit of frequency discrimina-

FIGURE 2.12. Schematic representation of the frequency (heavy
lines) corresponding to the tone sensations evoked by the
superposition of two pure tones of nearby frequencies
f_1 and $f_2 = f_1 + \Delta f$.

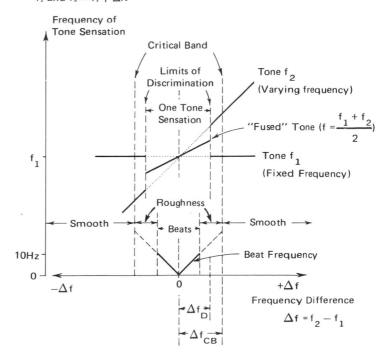

tion of f_1. Notice the extension of the critical band on either side of the unison ($\Delta f = 0$). We must emphasize again that this transition from roughness to smoothness is not at all sudden, as one might be tempted to conclude from Fig. 2.12, but rather gradual.

The limit for pitch discrimination and the critical band depend strongly on the average frequency $(f_1 + f_2)/2$ of the two tones (called the *center frequency* of the two-tone stimulus.) They are relatively independent of amplitude, but may vary considerably from individual to individual. The critical band is related to several other psychoacoustical phenomena, and there are many (and, as a matter of fact, far more precise) ways to define it experimentally (Section 3.4). Figure 2.13 shows the dependence of pitch discrimination Δf_D (Plomp 1964) and critical band Δf_{CB} (Zwicker, Flottorp, and Stevens 1957) with the center frequency of the component tones. For reference, the frequency differences that correspond to the musical intervals of a semitone, a whole tone, and a minor third are shown with broken lines. For instance, two tones in the neighborhood of 2,000 Hz must be at least 200 Hz apart to be discrimi-

FIGURE 2.13. Critical bandwidth Δf_{CB} (after Zwicker, Flottorp, and Stevens 1957) and limit of frequency discrimination Δf_D (Plomp 1964) as a function of the center frequency of a two-tone stimulus (linear scales). The frequency difference corresponding to three musical intervals is shown for comparison.

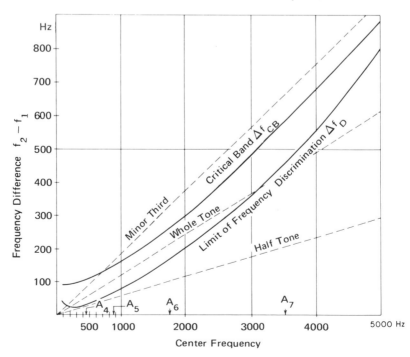

nated, and more than 300 Hz apart to sound "smoothly." Note the remarkable fact that the limit of pitch discrimination is larger than a half-tone,[5] and even larger than a while tone at both extremes of very high and very low frequencies. Note also the interesting fact that in the high pitch range, the critical band lies between the frequency difference that corresponds to a whole tone interval (qualified as a "dissonance") and that of a minor third (termed a "consonance")—i.e., roughly *extending over one-third of an octave*. In the lower frequency range there is an important departure: frequency discrimination and critical band are larger than a minor third (and even a major third). This is why thirds in general are not used in the deep bass register.

Compare Fig. 2.13 with Fig. 2.9: the limit for frequency discrimination Δf_D is roughly 30 times larger than the *jnd* for frequency resolution. In other words, we can detect very minute frequency changes of one *single* pure tone, but it takes an appreciable frequency difference between *two* pure tones sounding simultaneously, to hear out each component separately.[6]

What are the implications of these results for the theory of hearing? The existence of a finite limit for tone discrimination is an indication that *the activated region on the basilar membrane corresponding to a pure tone must have a finite spatial extension*. Otherwise, if it were perfectly "sharp," two superposed tones would always be heard as

---

[5] This may come as a surprise to musicians: they will claim that they can hear out very well the two component tones when a minor second is played on musical instruments! The point is this: the results shown in Fig. 2.13 only apply to *pure* tone superpositions, sounding *steadily* with constant intensity. When a musical interval is played with *real* instruments, the tones are not simple tones, they do not sound steadily, and a stereo effect is present. All this gives additional cues to the auditory system that are efficiently used for tone discrimination.

[6] There is an equivalent experiment that can be performed with the sense of *touch* to point out the difference between "resolution" and "discrimination." Ask somebody to touch the skin of your underarm for about one second on a fixed point with a pointed pencil *while you look away*. Then ask the person to repeat this at gradually displaced positions. It will require a certain small but finite minimum distance before you can tell that the position of touch has changed—this is the *jnd* for localization of a single touch sensation, or "touch resolution." Now ask the person to use two pencils and determine how far both touching points must be from each other before you can identify *two* touch sensations. This is the minimum distance for "touch discrimination," which turns out to be considerably larger than the *jnd*. Both touch resolution and discrimination vary along the different parts of the body. The equivalence between touch and hearing experiments is not at all casual: the basilar membrane is, from the point of view of biological evolution, a piece of skin with an enormously magnified "touch" sensitivity! This analogy has been profusely used by von Békésy (1960) in his superb experiments.

two separate tones as long as their frequencies differed from each other—no matter how small that difference—and no beat sensation would ever arise. Actually, the fact that the roughness sensation persists even beyond the discrimination limit, is an indication that the two activated regions still overlap or interact to a certain degree, at least until the critical band frequency difference is reached. An illustrative experiment is the following: feeding each one of the two tones f_1 and f_2 *dichotically* into a different ear, the primary beat or roughness sensation disappears at once, both tones can be discriminated even if the frequency difference is way below Δf_D, and their combined effect sounds "smooth" at all times! The moment we switch back to a monaural input, the beats or roughness come back. Of course, what happens in the dichotic case is that there is only *one* activated region on each basilar membrane with no chance for overlapping signals in the cochlea;[7] hence no beats or roughness.

At this stage the reader may wonder: if the region activated on the basilar membrane by *one* pure tone of *one* frequency is spatially spread, covering a certain finite range Δx along the membrane, how come we hear only *one* pitch and not a whole "smear" over all those pitches that would correspond to the different positions within Δx that have been activated? Unfortunately, we must defer the answer to the next chapter. Let us just anticipate here that a so-called *"sharpening"* process takes place after the signals from the stimulated region on the basilar membrane enter the neural network, in which the activity collected along the whole region Δx is "focused" or "funneled" into a more limited number of responding neurons while the surrounding neurons have been inhibited (contrast enhancement).

The beat phenomenon plays an important role in music. Whenever beats occur, they are processed by the brain giving us sensations that may range from displeasing or irritating to pleasing or soothing, depending on the beat frequency and the musical circumstances under which they appear. The peculiar, displeasing sound of an instrument out-of-tune with the accompaniment is caused by beats. The ugly sound of out-of-tune strings in a mediocre high-school orchestra is ugly in part because of the beats, and the "funny" sound of a saloon piano is caused by beats between out-of-tune pairs or triplets of strings in the middle and upper register. The fact that beats disappear completely when two tones have exactly the same frequency $(f_2 = f_1)$

[7] There is, however, an overlap of *neural signals* in the upper stages of the neural pathway, giving rise to "second order" effects, to be discussed in Sections 2.6–2.9.

plays a key role for the process of *tuning* an instrument. If we want to adjust the frequency of a given tone to be exactly equal to the frequency of a given standard (e.g., a tuning fork) we do this by listening to beats and "zeroing in" the frequency until the beats have completely disappeared.[8]

The critical band, too, plays a key role in the perception of music. We shall discuss this concept in later sections in more detail. For the time being, let us just remark that the critical band represents a sort of "information collection and integration unit" on the basilar membrane. The experimental fact that the critical band frequency extension Δf_{CB} is roughly independent of sound amplitude or loudness is a strong indication that it must be related to some inherent property of the structure of the sensorial organ on the basilar membrane, rather than to the wave form in the cochlear fluid. One may convert the frequency extension Δf_{CB} shown in Fig. 2.13 into spatial extension along the basilar membrane by using Fig. 2.8: one obtains an almost constant value of about 1.2 mm for the critical band. Another, even more significant correlation is the following: the critical band turns out to correspond to an extension on the basilar membrane "serviced" by the roughly constant number of about 1,300 receptor cells (out of a total of about 30,000 on the membrane) (Zwislocki 1965), independently of the particular center frequency (i.e., position on the membrane) involved.

A complex auditory stimulus (e.g., from two pure tones) whose components are spread over a frequency extension that lies *within* the critical band causes a subjective sensation (e.g., roughness, in our example) that usually is quite different from a case in which the extension *exceeds* that of the critical band (smoothness in the two-tone example). This is true for a variety of phenomena. It plays an important role in the perception of tone quality (Section 4.8). Its effect on pure tone superposition provides the basis for a theory of consonance and dissonance of musical intervals (Section 5.2).

2.5 Other First Order Effects: Combination Tones and Aural Harmonics

So far we have been analyzing the superposition effects of two pure tones whose frequencies were not too different from each other (Fig. 2.12). What happens with our tone sensations when the frequency of the variable tone, f_2, increases beyond the critical band, while f_1 is kept constant? The ensuing effects may be classified into two categories, depending on whether they originate in the ear or in the neural system, respectively. In this section we shall

[8] In this chapter we have discussed the case of beats between *pure* tones only. As we shall see later, they similarly occur for the complex tones of real musical instruments.

focus on a phenomenon belonging to the first category, the perception of *combination tones*. These tones are additional pitch sensations that appear when two pure tones of frequencies f_1 and f_2 are sounded together; they are most easily perceived if the latter are of high intensity level. These additional pitch sensations correspond to frequencies that differ from both f_1 and f_2, as can be established easily with pitch-matching or pitch-cancellation experiments (Goldstein 1970). *They are not present in the original sound stimulus*—they appear as the result of a so-called nonlinear distortion of the acoustical signal in the ear.

Let us repeat the experiment corresponding to Fig. 2.12 concerning the superposition of two pure tones, but this time turning up the loudness considerably and sweeping the frequency f_2 slowly up and down between the unison f_1 and the octave, of frequency $2f_1$. While we do this, we pay careful attention to the evoked pitch sensations. Of course, we will hear both the tone of constant pitch corresponding to the frequency f_1 and the variable tone f_2. But in addition, one clearly picks up one or more lower pitch tones sweeping up and down, depending on how we vary the frequency f_2. In particular, when f_2 sweeps upward away from f_1, we hear a tone of rising pitch starting from very deep. When f_2 sweeps downward starting at the octave $f_2 = 2f_1$, we also hear a tone of rising pitch starting from very deep. And paying even more attention, more than one low pitch tone may be heard at the same time. These tones, which do not exist at all in the original sound, are the combination tones.

The perhaps most easily identifiable combination tone at *high* intensity level is one whose frequency is given by the difference of the component frequencies,

$$f_{C1} = f_2 - f_1 \tag{2.4}$$

It is also called the *difference tone*. Notice that for values of f_2 very close to f_1, f_{C1} is nothing but the beat frequency (2.3). f_{C1} must be at least 20–30 Hz to be heard as a tone. As f_2 rises, f_{C1} increases, too. When f_2 is an octave above f_1, $f_{C1} = 2f_1 - f_1 = f_1$ —i.e., the difference tone coincides with the lower component f_1. When f_2 is halfway between f_1 and $2f_1$ —i.e., $f_2 = \frac{3}{2}f_1$ (a musical interval called the *fifth*)—the difference tone has a frequency $f_{C1} = \frac{3}{2}f_1 - f_1 = \frac{1}{2}f_1$ corresponding to a pitch one octave below that of f_1.

The two other combination tones that are most easily identified (Plomp 1965), *even at low intensity levels* of the original tones, correspond to the frquencies

$$f_{C2} = 2f_1 - f_2 \tag{2.5}$$
$$f_{C3} = 3f_1 - 2f_2 \tag{2.6}$$

Both tones f_{C2} and f_{C3} decrease in pitch when f_2 increases from unison toward the fifth, and they are most easily heard when f_2 lies between about $1.1f_1$ and $1.3f_1$. At high intensity of the original tones they also can be perceived quite well as low pitch sensations near the octave and the fifth, respectively. Note that the tones f_{C2} and f_{C1} coincide in frequency ($= \frac{1}{2}f_1$) when f_2 is at the fifth $\frac{3}{2}f_1$. In Fig. 2.14 we summarize the first order tone sensations evoked by the superposition of two pure tones of frequency f_1 and f_2. Notice that Fig. 2.12 is nothing but a "close-up" picture of what happens when the frequency f_2 is very near f_1 (hatched area in Fig. 2.14). The portions of the combination tones shown with heavier track are the ones easiest to be heard out (the actual extent depends on intensity).

How are these extra tone sensations generated? As pointed out above, they are *not* present in the original sound vibration of the eardrum. Careful experiments conducted on animals have shown that combination tone frequencies are not even present at the entrance of the cochlea (oval window membrane, Fig. 2.6); on the other hand, from direct neural pulse measurements (Goldstein 1970) one must conclude that there *are* indeed activated regions on the basilar membrane at the positions corresponding to the frequencies of the combination tones. They

FIGURE 2.14. Frequencies of the combination tones f_{C1}, f_{C2}, f_{C3}, evoked by a two-tone superposition (f_1, f_2). Heavy lines: most easily detected ranges of combination tones.

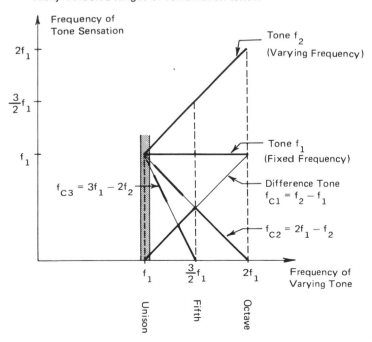

are thought to be caused by a "nonlinear" distortion of the primary waveform stimulus in the cochlea. It can be shown mathematically that, indeed, when two harmonic (sinusoidal) oscillations of different frequencies f_1 and f_2 enter a transducer of distorted (nonlinear) response, the output will contain, in addition to the original frequencies f_1 and f_2, all linear combinations of the type $f_2 - f_1$, $2f_1 - f_2$, $3f_1 - 2f_2$, $f_2 + f_1$, $2f_2 + f_1$, etc. More recent experiments (Smoorenburg 1972) indicate, however, that the difference tone (2.4) and the two other combination tones (2.5, 6) must originate in mutually independent cochlear mechanisms, respectively. The intensity threshold for the generation of the former is considerably higher than that of the latter, and roughly independent of the frequency ratio f_2/f_1. On the other hand, the intensity of the latter increases as f_2/f_1 approaches unity.[9]

It is interesting to note that even one *single* tone of frequency f_1 will give rise to additional pitch sensations when it is very loud. These additional tones, called *aural harmonics*, correspond to frequencies that are integer multiples of the original frequency: $2f_1$, $3f_1$, $4f_1$, etc.

Although all experiments pertinent to this section are most appropriately done with electronic tone generators, it is possible, at least qualitatively, to explore combination tones and aural harmonics using some musical instruments capable of emitting steady sounds at high intensity level. As a matter of fact, perhaps the most adequate "instrument" for this purpose is a dog whistle whose (high) pitch can be varied. A simple homemade experiment on combination tones can be done by blowing two such whistles at the same time—one at constant pitch, the other one sweeping the frequency away from and back to unison—and listening to low pitch tone sensations.

"Fake" combination tones can be easily generated in electronic organs and low quality hi-fi amplifiers and speakers. In these cases it is a non-linear distortion in the electronic circuitry and mechanical system of the speaker that generates these parasitical frequencies. In particular, the difference tone can be produced and listened to quite clearly with an electronic organ: turn the loudness up, pull 8' flute stops, play back and forth the sequence shown on the upper staff of Fig. 2.15, and listen for the low pitch tones indicated on the bottom staff.

Some of the difference tones thus generated are out of tune because of the equal-temperament tuning of the instrument (Section 5.3). We must point out again that what is heard in this experiment is a *fake* combination tone, in the

---

[9] The reasons for this difference in behavior are not yet understood. Nor is it known why the combinations $f_1 + f_2$, $2f_1 + f_2$, etc., do not appear as tone sensations.

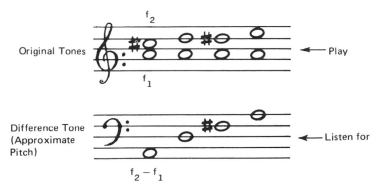

f_2

Original Tones ← Play

f_1

Difference Tone
(Approximate
Pitch) ← Listen for

$f_2 - f_1$

FIGURE 2.15

sense that the low pitch sensation is generated in the speaker and *not* in the ear. It is quite clear from this example why electronic circuitry and speakers of hi-fi systems and electronic organs should have a "very" linear response.

2.6 Second Order Effects: Beats of Mistuned Consonances

We now repeat the experiment of the preceding section with two electronically generated pure tones, but this time ignoring possible combination tone sensations. We feed both tones at low intensity level into the same ear; tone f_1 is held at constant frequency, whereas f_2 can again be varied at liberty. The amplitudes of both pure tones are held constant throughout the experiment. When we sweep f_2 slowly upward, we notice something peculiar when we pass through the neighborhood of the octave $f_2 = 2f_1$: a distinct beating sensation, quite different from the "first order" beats near unison, but clearly noticeable. When f_2 is exactly equal to $2f_1$, this beat sensation disappears. It reappears as soon as we mistune the octave, i.e., when f_2 is set $f_2 = 2f_1 + \varepsilon$, where ε (epsilon) represents only a few Hertz. The beat frequency turns out to be equal to ε. It is difficult to describe "what" is beating. Quite definitely, it is neither loudness nor pitch that is oscillating. Most people describe it is a beat in tone "quality." We call these *second order* beats; some prefer the name "subjective beats." They are the result of neural processing.

It is instructive to watch the vibration pattern on the oscilloscope while one listens to second order beats. This pattern is seen to change in exact synchronism with the beat sensations. Obviously, our auditory system must somehow be able to detect these changes of the *form* of a vibration pattern. Figure 2.16 shows several vibration patterns corresponding to the superposition of a fundamental tone of frequency f_1 and its octave $f_2 = 2f_1$ (of smaller amplitude), for four different values of the phase difference. As long as the octave is perfectly in tune, the phase difference remains

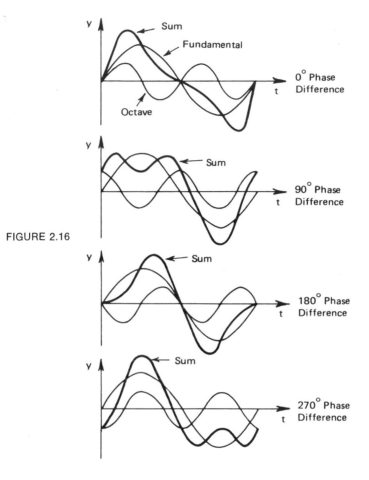

FIGURE 2.16

constant and the image on the oscilloscope static; any of the four superpositions sounds like the other—our ear does not distinguish one case from another. We say that the ear is insensitive to a static phase difference. But when we throw f_2 slightly out of tune: $f_2 = 2f_1 + \varepsilon$, the mutual phase relationship will change continuously with time and the resulting vibration pattern will gradually undergo a shift from one of the forms shown in Fig. 2.16 to the next. It can be shown mathematically that this cycle of changing vibration pattern repeats with a frequency ε, the amount by which the upper tone is out-of-tune from the octave. This obviously means that the ear is sensitive to a slowly changing phase difference between two tones. An equivalent statement is: *the auditory system is capable of detecting cyclic changes in vibration pattern forms.* Notice carefully that there is no macroscopic change in amplitude from pattern to pattern in Fig. 2.16—quite contrary to what happens with first order beats, which are cyclic changes of vi-

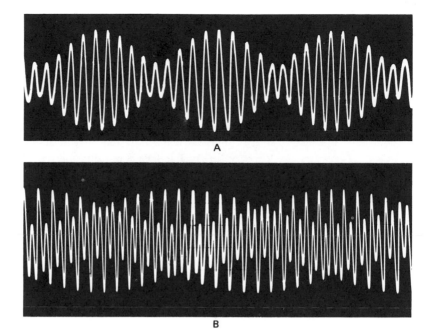

FIGURE 2.17. Comparison of first order and second order beats. (A) First order beats (mistuned unison); amplitude modulation with no change in vibration pattern form. (B) Second order beats (mistuned octave); pattern modulation with no change in total amplitude.

bration pattern amplitude (Fig. 2.11). Figure 2.17 shows two actual oscilloscope pictures confronting first order beats near unison and second order beats of a mistuned octave Note the amplitude modulation of the former and the vibration pattern modulation in the latter.[10] It is important to note that the second order beat sensation only appears in the low frequency range of the original two-tone stimulus. When f_1 (and f_2) exceeds about 1,500 Hz, second order beats cannot be perceived anymore (Plomp 1967).

Now we turn again to our experimental setup and explore the whole frequency range between unison and the octave. We discover that there are other pairs of values for f_2 and f_1, i.e., other musical intervals, in whose neighborhood beat sensations appear, though much less perceptible than for the octave. Two such "beat holes," as we may call them, can be found centered at the frequencies $f_2 = \frac{3}{2}f_1$

[10] A small modulation of the average energy flow does occur, at the rate ε (Rigden 1974). It is doubtful, however, that this fact can explain satisfactorily all observed psychoacoustical effects of second order beats.

and $f_2 = \frac{4}{3}f_1$ corresponding to the musical intervals fifth and fourth, respectively. Again watching the vibration pattern on the oscilloscope, at the same time as we listen, we realize that for a mistuned fifth $(f_2 = \frac{3}{2}f_1 + \varepsilon)$ and a mistuned fourth $(f_2 = \frac{4}{3}f_1 + \varepsilon)$, the vibration pattern form is not static (as happens with an exact fifth or fourth, i.e., for $\varepsilon = 0$), but changes periodically in form (not in amplitude). The second order beats of the fifth are "faster" than those of the octave (for the fifth, the beat frequency is $f_B = 2\varepsilon$, for the fourth $f_B = 3\varepsilon$). This is not the only reason why they are more difficult to pick up: the vibration pattern itself gets more and more complicated (i.e., departs more and more from that of a simple harmonic motion) as we go from the octave (Fig. 2.16) to the fifth and to the fourth. The more complex a vibration pattern, the more difficult it is for the auditory system to detect its periodic change (Plomp 1967).

There is an optimal relationship between the intensities of the component tones for which second order beats are most pronounced, which always places the *higher pitch tone at a lower intensity* (Plomp 1967). Finally, it is important to note that second order beats are also perceived when each of the component tones is fed separately *into a different ear*. In that case, a strange sensation of spatial "rotation" of the sound image "inside" the head is experienced (Section 2.9).

Second order beats of mistuned consonances of *pure* tones do not play an important role in music (mainly because pure tones don't). But they are an important ingredient to the understanding of the processing mechanism of musical sound (Section 2.8).

2.7 Fundamental Tracking

We now introduce another series of psychoacoustical experiments that have been of crucial importance for the theories of auditory perception. Let us consider two pure tones, a perfect fifth apart, of frequency f_1 and $f_2 = \frac{3}{2}f_1$. Figure 2.18 shows the resulting vibration (sum) for one particular phase relationship. Note that the pattern repeats the exact shape after a time τ_0, which is twice as long as τ_1, the period of the lower pitch tone. This means that the *repetition rate* $f_0 = 1/\tau_0$ of the vibration pattern of a fifth is equal to one-half the frequency of the lower tone:

$$f_0 = \frac{1}{2}f_1 \tag{2.7a}$$

We call this repetition rate the "fundamental frequency" of the vibration pattern. In this case, it lies one octave below f_1. If we consider two tones forming a fourth $(f_2 = \frac{4}{3}f_1)$, we can plot the resulting vibration pattern in the same way as was done for the fifth. The resulting repetition rate is now

$$f_0 = \frac{1}{3}f_1 \tag{2.7b}$$

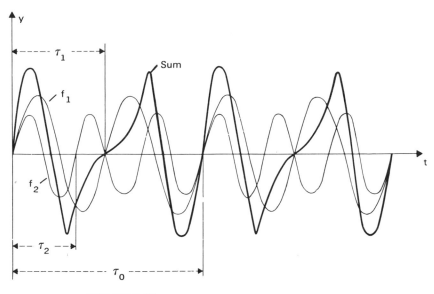

FIGURE 2.18

i.e., a twelfth below the lowest tone. For a major third
($f_2 = \frac{5}{4}f_1$), the repetition rate lies two octaves below f_1:

$$f_0 = \frac{1}{4}f_1 \qquad\qquad (2.7c)$$

*Our auditory system turns out to be sensitive to these repe-
tition rates.* Indeed, careful experiments have been performed in which the subjects were exposed to short sequences of stimuli made up of pairs of simultaneously
sounding pure tones, a fifth, a fourth, a third, etc., apart
(Houtsma and Goldstein 1972). These subjects were asked
to identify a *single* basic pitch of the "melody." Most of
them did indeed single out a pitch that is matched by a
frequency given by relations (2.7a), (2.7b), or (2.7c) respectively![11] It is important to point out that this pitch identification experiment demands that the two-tone complexes
be presented as a *time sequence* or melody. (When con-
fronted with a steady-sounding pair of pure tones, our auditory system fails to "seek" a single pitch sensation; it very
rapidly refocuses its attention in order to discriminate both
pure tone components as explained in Section 2.4.)

Note that the repetition rates (2.7a–c) of the above two-
tone complexes are identical to the frequencies of the cor-

[11] Please note that this experiment *must* be performed with pairs of
sinusoidal, electronically generated tones—it will *not* work on the
piano or on any other musical instrument. See, however, subsequent
remark on organs.

responding difference tones (e.g., see fourth, second, and first cases in Fig. 2.15). However, the experiments have shown that repetition rate detection is successful even if the intensities of the two tones f_1 and f_2 are low, way below the threshold for combination tone production. A difference tone (2.4) is thus ruled out (Plomp 1967a). Actually, repetition rate detection has been used in music for many centuries (and wrongly attributed to a combination tone effect). For instance, since the end of the sixteenth century, many organs include a stop (the "$5\frac{1}{3}$-foot fifth") composed of pipes sounding a fifth higher than the pitch of the written note actually played. The purpose is to stimulate or reinforce the bass *one octave below* (2.7a) the pitch of the written note (i.e., to reinforce the 16' sound of the organ). Of even older usage is the $10\frac{2}{3}$-foot fifth in the pedals, which, in combination with 16' stops, simulates or reinforces the 32' bass (two octaves below the written note).

The tone of frequency f_0 (2.7) is not present as an original component. This tone is called the *missing fundamental* (for reasons that will become apparent subsequently); the corresponding pitch sensation is called *periodicity pitch, subjective pitch, residue tone,* or *virtual pitch.* This pitch sensation should be clearly distinguished from the primary pitch of each one of the two original pure tones (also called spectral pitch). Experiments have shown that for "normal" loudness levels the frequency f_0 is *not* present in the cochlear fluid oscillations either (whereas combination tones are). Indeed, the region of the basilar membrane corresponding to the frequency f_0 (Fig. 2.8) may be saturated (masked) with a band of noise (sound of an infinite number of component frequencies lying within a given range) so that any additional excitation of that region would pass unnoticed—yet the missing fundamental will still be heard (Small 1970). Or one may introduce an extra tone slightly out of tune with f_0; first order beats should appear if the missing fundamental tone f_0 really did exist in the cochlea —yet no beat sensation is felt. A more drastic effect is that *the missing fundamental is perceived even if the two component tones are fed in dichotically,* one into each ear (Houtsma and Goldstein 1972). All of this indicates that the missing fundamental, or periodicity pitch, must be the result of neural processing at a higher level.

Subjective pitch detection, i.e., the capability of our auditory system to identify the repetition rate of the incoming, unanalyzed vibration pattern, only works in the lower (but musically most important) frequency range, below about 1,500 Hz. The more complex the vibration pattern, i.e., the smaller the interval between the component tones, the more difficult it is for the auditory system to identify the missing fundamental, and the more ambiguous subjective pitch becomes.

Let us now "turn around" relations (2.7) and find out which pairs of pure tone frequencies give rise to the *same* repetition rate or fundamental frequency f_0. We obtain

$2f_0$ and $3f_0$ $3f_0$ and $4f_0$

fifth   fourth

etc.

$4f_0$ and $5f_0$ $5f_0$ and $6f_0$

major third   minor third

In other words, if f_0 corresponds to the note shown on the lower staff in Fig. 2.19, the musical intervals shown in the upper staff yield that same note as a subjective pitch sensation. It is important to think of the notes in Fig. 2.19 as representing *pure* tones of just one frequency each, *not* as tones produced by real musical instruments. The individual components of frequency $2f_0$, $3f_0$, $4f_0$, $5f_0$,..., etc., are called the *upper harmonics* of the fundamental frequency f_0. Upper harmonic frequencies are integer multiples of the fundamental frequency. Any two successive tones of the upper harmonic series form a pair with the *same* repetition rate or fundamental frequency f_0. Therefore, all upper harmonics, if sounded together, will produce one single subjective pitch sensation corresponding to f_0—*even if that latter frequency is totally absent in the multitone stimulus!* This is the reason why in the above examples f_0 has also been called the "missing fundamental," and why the perception of this repetition rate is called *fundamental*

FIGURE 2.19. Two-tone stimuli (*upper staff*) that give rise to the same periodicity pitch (*lower staff*). The *B*-flat marked between parentheses must be tuned flat with respect to any scale in use (Section 5.3) in order to yield a *C* as periodicity pitch.

Two Tone Stimulus

Note Corresponding to
Periodicity Pitch

tracking. Note once more the striking property of this set of pure tones of frequencies $2f_0$, $3f_0$, $4f_0$, ... nf_0 ...—out of the infinite variety of possible pure tone superpositions, this is the only one whose components, taken in pairs of consecutive tones, yield one and the same repetition rate. Conversely, this is the reason why *any* periodic tone with a complex but repetitive vibration pattern (of repetition rate f_0) is made up of a superposition of pure tones of frequencies nf_0 ($n =$ whole number).

The above-mentioned psychoacoustical experiments with two-tone complexes have been extended to include melodies or sequences composed of *multitone* complexes starting on the nth harmonic (i.e., superpositions of pure tones of frequencies nf_0, $(n + 1)f_0$, $(n + 2)f_0$, etc.). Although, again, the tone of fundamental frequency is missing, the subjective pitch assigned to these tone complexes always corresponds to f_0. As a matter of fact, the more harmonics included, the clearer the periodicity pitch is heard (unless the starting harmonic order is very high, i.e., n is large). The most crucial pairs of neighboring harmonics for periodicity pitch determination are those around $n = 4$ (Ritsma 1967). Because "real" musical tones happen to be made up of a superposition of harmonics (Chapter 4), *fundamental tracking is the auditory mechanism that enables us to assign a unique pitch sensation to the complex tone of a musical instrument* (Plomp 1967a).[12]

2.8
Auditory Coding
in the Peripheral
Nervous System

The discovery of second-order effects in auditory processing, such as the perception of phase changes, beats of mistuned consonances, and fundamental tracking, has had a great impact on the theory of hearing. Indeed, these effects cannot be explained appropriately with a simple-minded "place theory" (Section 2.3). This does not mean, however, that this latter theory is wrong—it simply means that it has to be expanded or implemented. On one side, the perception of beats of mistuned consonances (Section 2.6) is an indication that the auditory system somehow does obtain and utilize information on the time structure of the acoustical vibration pattern. Fundamental tracking (Section 2.7), on the other hand could, in principle, imply *two* alternatives: (1) a mechanism performing detailed *time* pattern analysis, or (2) a central mechanism that analyzes the neural information on the complicated *spatial* excitation pattern elicited along the basilar membrane by a complex harmonic

---

[12] Perhaps the most convincing example of fundamental tracking of complex tones is given by the fact that one is still able to perceive the correct pitch of bass tones from a cheap transistor radio *in spite* of all frequencies below 100–150 Hz being cut off by the inadequate electronic circuitry and speaker!

tone, with the precise "instruction" to yield a single pitch sensation. This latter alternative could imply that the central nervous system must first *learn* this pitch assignment operation at some early stage of the development of an individual.

In order to understand the underlying mechanisms that have been proposed in recent years to explain these various psychoacoustical phenomena, it is necessary first to analyze some operational aspects of the auditory nervous system. We start with the arrangement of the actual receptor units on the basilar membrane. These sensor units or *hair cells* are grouped in "internal" and "external" rows that run along the basilar membrane from the base to the apex (Fig. 2.7). Nerve endings making contact with the hair cells receive orders from the latter to fire electrical impulses whenever the mechanical stimulus (bending) of the hair cell surpasses a given threshold. A most significant feature is the arrangement of these nerve endings. Whereas a single nerve fiber usually contacts only 2–4 neighboring sensor units of the *inner* row of hair cells, thereby receiving messages from a spatially very limited region, single nerve fibers innervating the *outer* rows make contact with sensor units that may lie several millimeters from each other (Davis 1962), thus being able to collect simultaneous information from a wider range of frequencies. These outer hair cells seem to be more sensitive to a given acoustic stimulus (lower threshold) than are those of the inner row.[13] Furthermore, neurons wired to the hair cells of the inner rows seem to respond mainly to the *velocity* of the basilar membrane motion, whereas outer-row hair cells signal according to *displacement* (Dallos et al. 1972). All of this suggests that both groups of hair cells play quite distinct roles in auditory detection (see also pp. 47 and 48.) It is significant that only higher mammals possess *two* types of rows of hair cells.

The fundamental processing and transmission unit of the neural system is the nerve cell or *neuron* (e.g., Eccles 1970). A typical neuron is shown in Fig. 2.20. One distinguishes the cell *body* or soma, a series of ramified processes called *dendrites*, and an elongated fiber, the *axon*, which also may split into multiple branches. The dendrites and the cell body are the *receptors* of incoming neural signals, and the axon is the *effector*, which passes signals on to other neurons. These neural signals consist of electrical impulses (electric voltage variations) of several tens of millivolts. These impulses can be registered by implanting mic-

---

[13] Recent studies seem to indicate, however, that the great majority of sensory neurons is associated with the inner hair cells (Spoendlin 1970).

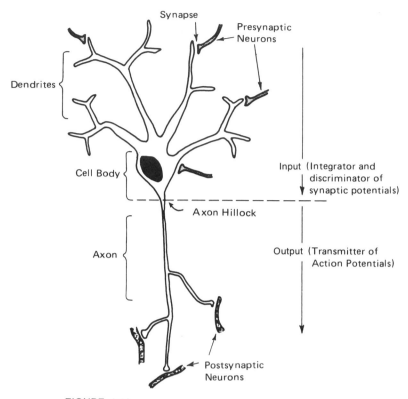

FIGURE 2.20

roelectrodes into the neuron (a procedure that does not impair the normal function of the cell). In the axon each impulse, called action potential, has nearly the same shape and duration (several tenths of milliseconds), and travels from the cell body (the axon hillock) to the axon endings. The action potential represents the fundamental, elementary neural output message. An "integrated" neural message is given by the *rate* or the *distribution in time* with which individual impulses are fired along the axon.

The axon is "wired" to dendrites or cell bodies of other neurons. The contact points are called *synapses*. One given axon may be in synaptic contact with many other cells; conversely, one given cell may be wired to incoming axons from hundreds or thousands of other cells.[14] An impulse arriving at a synaptic contact causes the release of a chemical substance from the presynaptic cell into the space between both cell membranes (synaptic cleft). The presence of this substance triggers an electrical impulse in the post-

---

[14] Exceptions are the unipolar neurons the single dendrites of which form the afferent nerve fibers.

synaptic cell, the "synaptic potential." Synaptic potentials are of varying shape and of longer duration than the standard-sized action potentials that propagate through the axon. There are two distinct types of synapses, *excitatory* and *inhibitory*, which evoke synaptic potentials of mutually opposite polarity. If, within a given short interval of time, a neuron receives a number of excitatory stimulations that exceed the number of simultaneously arriving inhibitory signals by a certain *threshold* value, it will respond by firing an impulse through its axon. Otherwise it will remain silent. We conclude that the dendrites and the cell body function as the information collection and integration system of the neuron, thus representing the fundamental information-processing unit in the nervous system. It is important to note that whether a neuron will fire an output signal is determined by both the spatial *and* temporal distributions of incoming signals from presynaptic neurons. A single neuron may only impart either excitatory or inhibitory orders to other neurons. When an inhibitory neuron fires a pulse to another inhibitory neuron it *cancels* the inhibitory effect of the latter. There are indications that the response of neurons wired to the inner-row hair cells is excitatory, whereas that from the outer row is inhibitory (Sokolich and Zwislocki 1974).

A characteristic time delay (typically less than a millisecond) occurs between the arrival of an impulse to a synapse and the formation of the response in the postsynaptic cell. This time delay makes it possible that the neural activation triggered by a single external stimulus may subsist or "reverberate" for a considerable amount of time when it propagates through a serial relay of thousands of successive synaptic steps in the cerebral tissue. Such "reverberation" may be a key process in the buildup of time-dependent patterns of neural activity (e.g., Section 4.9). After each activation, a neuron has a refractory period during which it cannot be re-excited, or during which its firing threshold is increased. One single neuron is just a "microscopic" component of a conglomerate of about 10 billion neurons in the human brain. It is in the "wiring scheme" among these 10 billion neurons that the "secret" of intelligence, behavior, feelings, and fears of human beings lies buried. In a broad macroscopic sense, the development of this wiring scheme is commanded by the genetic code of the species. However, in an important part of the neural system, especially in the cerebral cortex, the particular distribution of active synaptic contacts is the result of repetitive action of stimulation patterns—in other words, of experience and learning.

When an individual neuron of the neural tract pertaining to a given sense organ is observed with a microelectrode, one generally finds a correlation between the firing rate and the magnitude of certain physical parameters of the original stimulus. Individual firings usually do not occur equally spaced at regular time intervals. It is either the *fact* that a neuron is firing or the *average rate* that counts (however, see discussion further below). A neuron may be found to fire even in the absence of the original stimulus; this is called spontaneous activity and may happen at average rates up to several tens of Hertz in the peripheral neurons. For such a neuron it is the *change* in firing rate (increase *or* inhibition) that constitutes the neural message. In general a persisting constant stimulus will evoke a firing rate that gradually decreases with time until it levels off at a certain lower value. This phenomenon is called *adaptation*. However, some neurons are found to respond only to *time* changes of stimuli, others to a whole complex of particular spatial configurations or time patterns of the stimulus (*feature detectors*). As a general rule, the higher up we move along the neural pathway from the receptors to the brain cortex, the more complex and elaborate the stimulus features to which a given neuron responds (see Fig. 2.25).

We are now in a better position to discuss how the neural system may collect and code information on acoustical vibration patterns. When the acoustical signal of a single pure tone of given frequency arrives at the ear, the basilar membrane oscillations stimulate the hair cells that lie in the resonance region corresponding to that frequency (Section 2.3). It has been found by implanting microelectrodes in acoustically activated cochlear nerve fibers, that a given fiber has a lowest firing threshold for that acoustical frequency f which evokes a maximum oscillation at the place x of the basilar membrane (Fig. 2.8) innervated by that fiber (this frequency for maximum response has been called the fiber's "best frequency"). Turning now to the actual distribution in time of individual pulses, recent measurements (Zwislocki and Sokolich 1973) have shown that maximum firing rate is associated with maximum *velocity* of the basilar membrane *when it is moving toward the scala tympani*; inhibition of the firing rate occurs during motion in the opposite direction, toward the scala vestibuli. Moreover, the *momentaneous position* of the basilar membrane has a (less pronounced) excitatory or inhibitory effect, depending on whether the membrane is momentarily distorted toward the scala tympani, or away from it, respectively. Both effects add up to determine the total response. Figure 2.21 shows a hypothetical time distribution of neural impulses in a nerve fiber of the inner ear connected to the appropriate resonance region of the basilar membrane, when it is excited by a low-frequency tone of a trapezoidal vibration pattern shape (after Zwislocki and Sokolich 1973).

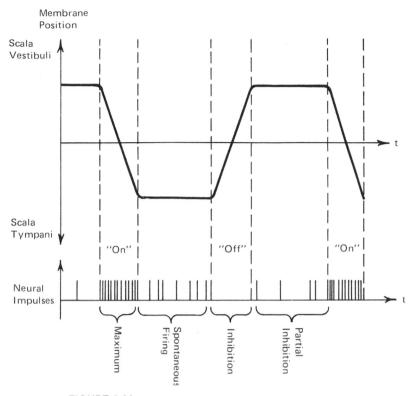

FIGURE 2.21

Close inspection of this figure reveals how information on repetition rate (actually, the repetition period) of the original acoustical signal can be coded in the form of "trains" of nerve impulses. Figure 2.21 would correspond to an ideal case of low frequency. Actually, the acoustical frequencies are usually higher than those of neural firing rates, and the *real* situation corresponds rather to one in which the "on" and "off" intervals are much less clearly delineated because of their short duration (as compared to the refractory period of a typical neuron) and because of the random character of the impulse distribution. The only *statistically* important property is that there will be more impulses falling into "on" intervals than into "off" intervals. As a result, for pure tones, *the time interval between successive impulses will tend to be an integer multiple of the sound vibration period* τ (Kiang et al. 1965). It is clear that the higher the frequency of the tone, the less well-defined this grouping will be. For frequencies above a few thousand Hertz it does not work at all. When several fibers receiving stimuli from a narrow region of the basilar membrane are bundled together (as occurs in the auditory nerve), the sum of their impulses (as detected by a macroelectrode that simultaneously makes contact with many fibers at the same time) will appear in synchrony with the auditory stimulus. These collective synchronous nerve signals have been called *volleys*.

It follows from the preceding section that a given neural fiber of the auditory nerve is capable of carrying two types of information.

1. The simple fact that *it is firing* tells the auditory system that the basilar membrane has been activated at or near the region innervated by that fiber—the spatial distribution (or "tonotopic" organization) of firing fibers codes the information on primary pitch ("place theory" of hearing). This process works for the whole range of frequencies.
2. The actual *time distribution* of impulses carries information on repetition rate or periodicity and possibly also on details of the vibration pattern itself (see below). This only works in the lower frequency range.

There is no doubt that information as to the *place* of excitation is used by the auditory system at all levels. But does this system actually utilize the information contained in the *time* distribution of neural pulses schematically shown in Fig. 2.21?

First, let us return for a moment to the perception of single, pure (sinusoidal) tones. Several arguments point to the fact that the time distribution of neural pulses is *not* utilized in the perception of the pitch of a *pure* tone. For instance, theoretical calculations (Siebert 1970) predict that if primary pitch were mediated by time cues, the *jnd* of frequency resolution (e.g., see Fig. 2.9) would be independent of frequency (which it is not), and in turn should decrease with increasing stimulus amplitude (which it does not).

That time cues are largely ignored in primary pitch perception of pure tones may not come as a surprise. But what about perception of beats of mistuned consonances and periodicity pitch of harmonic complexes? It is difficult to find an explanation of beats of mistuned consonances and other phase-sensitive effects without assuming that at some stage a mechanism analyzes the temporal fine structure of the vibration pattern of the stimulus. Indeed, we may invoke the effect shown in Fig. 2.21 to attempt an explanation of how information on the vibration pattern and its variations (second order beats) might be coded. Consider the superposition of two tones an octave apart. Assume that the resulting vibration pattern is that shown at the bottom of Fig. 2.16. Two resonance regions will arise on the basilar membrane, centered at positions x_1 and x_2 corresponding to the two component frequencies f_1 and $f_2 = 2f_1$ (Fig. 2.8). In the cochlear nerve bundle we shall have two main foci of activity centered at the fibers with "best frequencies" f_1 and f_2, leading to two primary pitch sensations, one octave apart. However, the resonance regions on the basilar membrane are rather broad, with sufficient overlap in the region between x_1 and x_2 where the points of the membrane will vibrate according to a superposed pattern related to the original motion of the eardrum.[15] Fibers connected to that overlapping region will thus respond with firings that are grouped in "on" intervals of enhanced rate which, say, correspond to the descending (negative slope) portions of the second graph in Fig. 2.16. Note that in this case the "on" intervals are not of equal duration but instead form an alternating "short–longer–short–longer" sequence. If the two tones were a fifth apart, the vibration pattern of the overlapping region could be that shown in Fig. 2.18, leading to a yet different type of sequence of "on" intervals. Periodicity of this sequence thus would represent the information on repetition rate, whereas structure of the sequence (a sort of "Morse code") would give information on the vibration pattern. Such a fine structure indeed has been identified statistically through electrophysiological measurements. Figure 2.22 is an example (a so-called histogram) of the distribution of time intervals between neural pulses in an auditory nerve fiber (Rose et al. 1969), for a stimulus corresponding to a musical fifth in a given phase relationship. Note the difference in the relative number of times (vertical axis) a given interval between successive pulses (horizontal axis) appears. This represents the (statistical) "Morse code" mentioned above, carrying information on the vibration pattern. The greater the complexity of the original vibration pattern and the higher the frequency of the component tones, the more "blurred" the information conveyed by the pulse sequence will be, i.e., the more difficult to be interpreted at the higher brain levels. Detailed analysis of the neural pulse time distribution would require the operation at some level of what is called a *temporal autocorrelation* mechanism (Licklider 1959), in which a pulse "train" is compared with previous pulse trains, whereby similar repetitive features are enhanced and all others (nonperiodic) are suppressed). However, there is no anatomic evidence as yet for the existence of such a mechanism in the afferent auditory nervous system.

Time cues are also operative in the mechanism responsi-

---

[15] Traveling waves in the cochlear fluid change their phase relationship as they propagate, thus altering the *actual* form of the vibration pattern at different points of the basilar membrane.

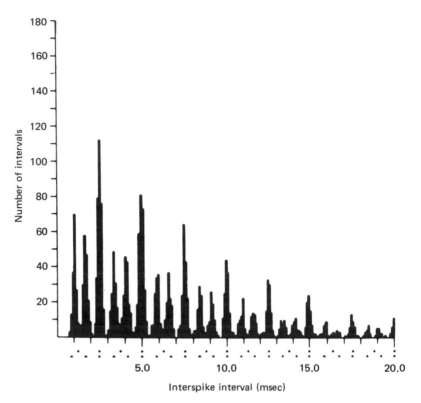

FIGURE 2.22. A histogram showing the number of times (vertical axis) a given interval between neural spikes occurs (horizontal axis), in an auditory nerve fiber stimulated with a two-tone superposition of a given phase relationship (fifth).

Rose et al. 1969. By permission from the authors.

ble for the sensation of spatial (stereo) *sound localization*[16] (e.g., Molino 1974). This should involve a process called temporal *crosscorrelation* of the neural signals from both cochleas in which the time difference between the signals from both cochleas is determined. There *is* physiological evidence that such a mechanism exists (in the medial superior olive, Fig. 2.25). A neural model of a crosscorrelator has been proposed by Licklider (1959). In this model (Fig. 2.23) it is assumed that an ascending neuron can fire only if it is excited simultaneously by both incoming fibers. Because a neural signal propagates with a finite velocity along a fiber, simultaneous arrival at a given ascending neuron ending requires a certain *time difference* between the

[16] Intensity cues (amplitude difference between the sound waves arriving at the two ears) also contribute to sound localization, especially at high frequencies.

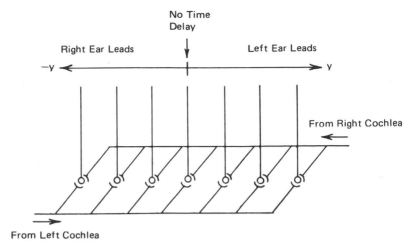

FIGURE 2.23. Model for a neural crosscorrelation mechanism (interaural time-difference detector).

After Licklider 1959.

original signals in both cochleas. For instance, exact simultaneity (zero time difference) of both cochlear signals would fire the ascending neuron located exactly in the center, because that is the place at which both right and left signals meet. If, however, the original signal is detected first in the *right ear*, its pulse will travel past the middle point until it meets the delayed pulse from the left ear. It is easy to see that the location y (Fig. 2.23) of the activated ascending neuron will depend on the interaural time delay, which in turn depends on the direction of the incoming sound. Two tones, a mistuned interval apart, fed into *separate* ears, may "foul up" the crosscorrelator: The gradually shifting phase difference between the two tones (e.g., Fig. 2.16) will be interpreted by this mechanism as a changing difference in *time of arrival* of the left and the right auditory signals, hence signaling to the brain the sensation of a (physically nonexistent) cyclically changing sound direction! This is why two pure tones forming a mistuned consonant interval, presented dichotically with headphones, give the eerie sensation of a sound image that seems to be "rotating inside the head" (p. 40).

The most controversial issue has been the question of whether or not time sequence analysis of neural pulses is a necessary hypothesis to explain periodicity pitch perception.[17] A temporal autocorrelation mechanism with its potential capability of detecting the repetition rate of neural signals, could indeed explain some important psychoacous-

[17]For a comprehensive review, see Wightman and Green (1974).

tical results of fundamental tracking (but *not all*). If it is not a time cue analysis, what *is* the mechanism that enables us to attach a single pitch to a harmonic tone complex—even if the fundamental is not present in the original stimulus? Why at all do we perceive the pitches corresponding to the frequencies given by relations (2.7a)–(2.7c) when a melody is played with the corresponding harmonic two-tone complexes?

The most recent ideas that may lead toward an explanation of these effects (e.g., Terhardt 1972, Wightman 1973, Goldstein 1973) are presented here in a highly oversimplified way. "Natural" sounds of human and animal acoustic communications contain an important proportion of *harmonic* tones (vowels, bird song, animal cries). Such tones share a common property—they are made up of a superposition of harmonics, of frequencies nf_1, integer multiples of a fundamental f_1 (p. 43). These tones elicit a complicated resonance pattern on the basilar membrane, with multiple amplitude peaks, one for each harmonic (Fig. 2.24).[18] In spite of its complexity this pattern does bear some invariant characteristics. *On such invariance is the particular distance relationship between neighboring resonance maxima.*[19] [For higher order harmonics (n greater than 7 or 8) this relationship loses its physical definition because of mutual resonance overlaps.] Based on the general principle of maximum economy, minimum redundancy of neural information processing, we either learn at an early age (Terhardt 1972, 1974), or we have a built-in mechanism (Wightman 1973, Goldstein 1973), to recognize this invariant characteristic as belonging to "one and the same thing." We shall call this mechanism of recognition the *central pitch processor*. The main function of this neural unit is to transform the peripheral activity pattern into another pattern in such a way that all stimuli with the same periodicity are similarly represented. The result is a unique pitch sensation—in spite of the many concurrent harmonics and the ensuing complexity of the primary excitation pattern. This unique pitch sensation corresponds to that of the fundamental component f_1, which in "natural" sounds is usually the most prominent one (intensity–wise). All of this should work much in analogy with visual pattern recognition. For instance, when you look at the symbol ш, it may not convey any "unique" meaning at all (your interpretation would probably depend upon the spatial orientation of the symbol and the context

---

[18] Because of the particular relationship between resonance place and frequency (Fig. 2.8), the resonance regions crowd closer and closer together as one moves up the harmonic series (as n increases.)

[19] Another invariant characteristic is the high coherency of the macroscopic time variations of this complicated excitation pattern over the whole spatial domain of the basilar membrane.

FIGURE 2.24

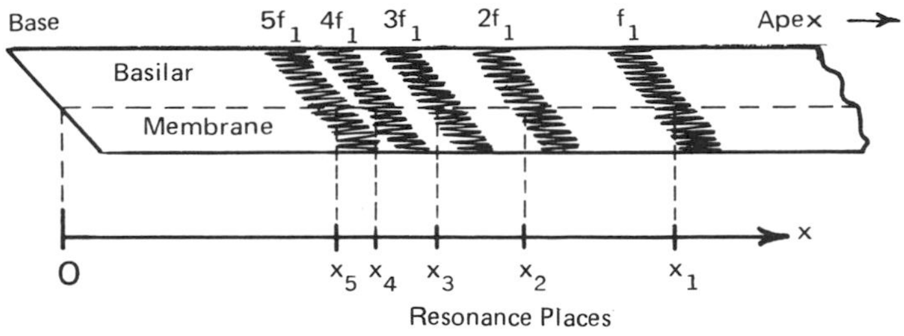

in which it is shown). But anyone familiar with the Cyrillic alphabet clearly perceives it as just "one thing" (the letter "shch"), no matter where in the visual field, and in what orientation, it is projected.

It is assumed that we have built into our central processing system basic *templates* with which to compare the complex structures of spatial excitation pattern from the basilar membrane. Whenever a match is achieved, a unique pitch sensation is elicited. This matching process works even if only a partial section of the excitation pattern is available. If instead of a "natural" complex sound, we are exposed to one in which some normally expected elements are suppressed (e.g., a missing fundamental), the partially truncated excitation pattern on the basilar membrane fed into the recognition mechanism of the pitch processor still may eventually be matched, within certain limitations. Again, we find many analogies in visual pattern recognition. A remarkable example is the apparent perception of nonexisting— but *expected*—contours, which we experience when we look at the letters of the title printed on the title page of this book (Terhardt 1974).

The above-described matching process works even if the harmonic components are fed alternatively, but simultaneously, into separate ears (e.g., Houtsma and Goldstein 1972). This obviously means that the central pitch processor must be located in some upper stage of the auditory pathway, after the input from both cochleas has been combined. Furthermore, the matching process works even if only *two* neighboring harmonics of a complex tone are presented (Section 2.8). In such a case, however, the matching mechanism may commit errors—and lock into one of *several* "acceptable" positions.

Three theories correctly predict many quantitative results of psychoacoustical measurements concerning periodicity pitch. One of them (Goldstein 1973) assumes that the neural information on the spatial positions of resonance maxima on the basilar membrane is not sharply defined, fluctuating statistically within certain bounds. The match of the "template" is expected to be one that statistically minimizes the differences with the real signal within the expected fluctuations (a maximum likelihood estimation of fundamental frequency and harmonic order—more on this in Section 4.8 and Appendix II). An almost equivalent theory (Wightman 1973) assumes the operation of an autocorrelation mechanism working in the spatial domain (in distinction to the temporal autocorrelation mentioned above. p. 51). The spatial activity distribution elicited in an ensemble of nerve fibers receiving information from both cochleas is fed into a neuronal network in which the activation of one given spatial area is compared quantitatively with that existing in distant areas, at any given time. In this process, certain input features are converted into one focus of output acivity whose spatial location encodes *one* quality (i.e., pitch) that is related to a given invariant characteristic of the input (i.e., the distance relationships between excitation maxima on the basilar membrane), regardless of other possible input variables (e.g., the intensities or the phases of the various harmonics). It should be pointed out that this "spatial" autocorrelation mechanism most likely does involve a learning process too. Indeed, conditioning may be required to make the system responsive to the most frequently occurring relevant spatial constellations of activity maxima (Fig. 2.24) elicited by natural (harmonic) sounds.[20] Neither of these two theories proposes *how* the key algorithms (template matching or spatial autocorrelation, respectively) are actually carried out by the pitch processor in the nervous system. Yet neuronal networks are believed to exist that are perfectly capable of performing the operations of neural pulse summation and multiplication required in the execution of these algorithms. Terhardt's theory (1974) comes closest to proposing a neural wiring scheme. Indeed, its computations are based on a *learning matrix,* an analog circuit that "learns to respond"[21] to characteristic features of the most frequently occurring input configurations (i.e., to the distance relationships between input excitation maxima caused by a complex tone). We shall return to these theories later, when we discuss explicitly the perception of complex musical tones (Section 4.8 and Appendix II), and consonance and dissonance (Section 5.2).

Finally, let it be stated that at the present time *one cannot exclude* completely the possibility that at least partial use is made of the time distribution of neural pulses, in the pitch perception of complex tones. It is hard to believe

[20] It is rather improbable that the required (quasilogarithmic and cochlear-dependent) "metric" is built in genetically.

[21] This is accomplished in the laboratory model by appropriately decreasing electrical resistances between transmission lines (the rows and columns of the matrix) that are simultaneously activated (conduce current) by a given, repeatedly presented, input configuration.

that the nervous system, always geared to work with such an astounding efficiency, with so many backup systems, is *not* taking advantage of the handy "Morse code" information (Fig. 2.21) that actually exists in peripheral auditory transmission channels! As a matter of fact, some psychoacoustic experiments seem to demand an explanation via time-cue analysis. For instance, low-frequency pure tones of very short duration (2–3 actual vibration cycles) *can* give rise to a clear pitch sensation (Moore 1973). Or if an acoustical signal (white noise) is presented to one ear, and the same signal is fed into the other ear delayed by an interval τ (a few milliseconds), a *faint pitch* corresponding to a frequency $1/\tau$ is perceived (Bilsen and Goldstein 1974). Neither of these results can be explained satisfactorily by a "place" theory (spatial cue analysis). Several more years of systematic research will have to be awaited before a more categoric picture can be drawn.

One thing is clear from the preceding discussion: subjective pitch perception requires that "higher order" operations of pitch extraction be performed in the central nervous system *after* the input from both cochleas has been combined. For this reason, we conclude this chapter with a summary description of some of the most relevant features of the auditory pathway (Whitfield 1967, Brodal 1969). This should also serve as a reference for discussions to come in later chapters. The anatomic exploration of neural pathways and their interconnections is an extremely difficult experimental task. Neurons are cells the processes of which (axons or dendrites) may be many centimeters long; each neuron, especially in brain tissue, may receive information from thousands of cells while it relays information to hundreds of others. It is practically impossible to follow the connection pattern microscopically even in the case of one cell. Only gross estimates of general features of the pathways can be made through the use of various techniques of cell staining, cell degeneration, or by following through the tissue the course of specific patterns of neural activity.

Figure 2.25 diagrams the auditory pathway from the cochlea to the auditory receiving area of the cerebral cortex in *flow chart* form. This flow chart depicts information–transmission channels and relay-processing stations, and bears no scale relationship whatsoever to the actual neuroarchitectonic picture. The *spiral ganglion* is the neural network in the cochlea, a first processing stage in this pathway. It is here that neurons contacting inner- and outer-row hair cells have a first chance to interact, determining the particular spatiotemporal distribution of activity in the *acoustic nerve* (the VIIIth cranial nerve), which conveys this information to the brain. The next processing stage, located in the medulla oblongata, comprises the *cochlear nuclei,*

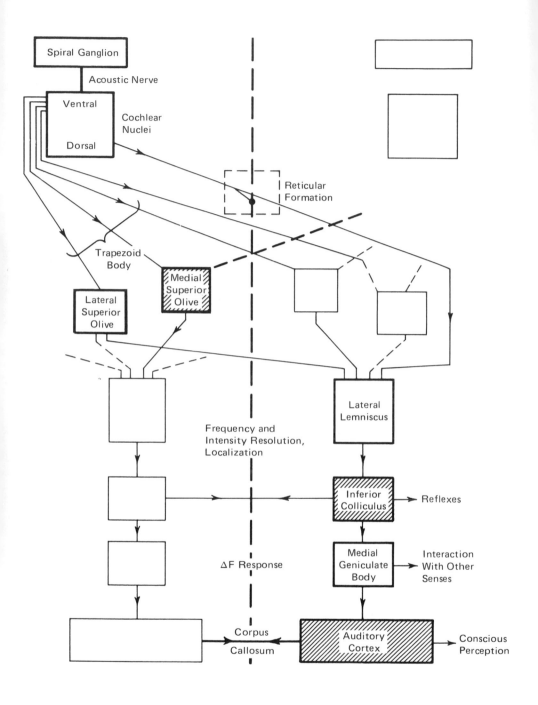

FIGURE 2.25. Flow chart of neural signals in the auditory pathway from one ear through the brain stem to the auditory cortices.

composed of three subdivisions the elaborate structure of which is responsible for the first steps of sound resolution and discrimination tasks. From here, neural information is channeled into three main bundles. One crosses over directly to the opposite *contralateral* side and enters the *lateral lemniscus*, the main channel through the brain stem (pons). Some fibers terminate in the *reticular formation*, a diffuse network in the brain stem that plays the role of a major cerebral "switchboard."[22] Another bundle (the *trapezoid body*) sends fibers from the ventral cochlear nucleus to important relay and processing stations, the *lateral* and *medial superior olives* (Fig. 2.25). Of these, the medial superior olive is the first intraaural signal-mixing center. This is the place at which a crosscorrelator (Fig. 2.23) may yield the information necessary for sound source localization. Finally, a third intermediate bundle leads from the ventral cochlear nucleus to the contralateral olivary complex.

The three upper stages involve the *inferior colliculus*, the *medial geniculate body*, and finally, the *auditory cortex* (Fig. 2.25). Some fibers are connected to the *superior colliculus*, which is also innervated by visual pathways (p. 146). Note the interconnections at these various stages with the contralateral pathway and with other sensory pathways and brain centers.

Not shown in Fig. 2.25 is a network of *efferent fibers*, which carries information from the upper stages down to lower ones and terminates in the cochlea. This system undoubtedly plays a role in the control of incoming afferent information, but it is not yet known in all of its details. The lower tract of the efferent network, the *olivocochlear bundle*, may participate in an important way in the sharpening process (pp. 32, 91).

Finally, let us point out some generalities that may be useful for later chapters. At the initial stage there is a very specific geometric correspondence between activated neural fibers and spatial position of the source stimulus on the basilar membrane. Indeed, the spatial distribution of excitation along the basilar membrane is mapped continuously into spatial distribution of neural activity across the bundles of fibers. This is particularly apparent in each one of the cochlear nuclei. As one moves up to higher stages, however, this correspondence is gradually lost (except in an anesthetized state). The neural response becomes increasingly representative of complex *features* of the sound signal, being more and more influenced by the behavioral state and

[22] This structure, which receives raw data from the senses and the body as well as elaborate information from the brain, is responsible for activating or inhibiting cerebral processing according to instantaneous needs, controls sleep and consciousness, and influences many visceral functions.

the performance of the individual. Contralateral (i.e., cross-ing) channels are "better" information carriers than are ipsilateral channels (to the same side)—if conflicting infor-mation is presented to both ears, the contralateral channel tends to override the information that is carried to a given hemisphere by the ipsilateral channel (Milner et al. 1968).

At the stage of the inferior colliculus, good resolution of frequency, intensity, and direction of sound already exists; so does a selective response to up–down frequency sweeps. Reflexes work, but at this stage there is no evidence for conscious perception of sound, as has been shown by abla-tion experiments. In the *medial geniculate body* (and prob-ably in the superior colliculus) some pattern recognition capability is already operational. At this stage information exists on *where* a given sound stimulus is and *where it is going* in space and time. The first integration with informa-tion from other senses takes place.

The last stage of incoming information processing is performed in the receiving area of the *auditory cortex*. The primary function is to identify the stimulus, to integrate it into the currently displayed image of the environment, and to make it available to the conscious state of the brain. Indeed, from here on, the information is distributed to other brain centers, where it is stored, analyzed, integrated into the whole brain function—or discarded as irrelevant. The *corpus callosum* (Fig. 2.25), a gigantic commissure of about 200 million fibers connecting both cerebral hemis-pheres, plays a key role in global information processing, especially in view of the remarkable specialization of the two hemispheres, as already mentioned in Section 1.5. We shall return to this subject in Section 5.6.

3 Sound Waves, Acoustical Energy, and the Perception of Loudness

In the preceding chapter we studied simple sound *vibrations* and their subjective effects, without investigating how they actually reach the ear. We referred to experiments in which the sound source (headphone) was placed very close to the eardrum. In this chapter we shall discuss the process of sound energy *propagation* from a distant source to the listener and analyze how this acoustical energy flux determines the sensation of loudness.

3.1 Elastic Waves, Force, Energy, and Power

When sound propagates through a medium, the points of the medium vibrate. If there is no sound at all and if there is no other kind of perturbation, each point of the medium will be at rest and remain so until we do something to the medium. The position in space of a given point of the medium when the latter is totally unperturbed is called the *equilibrium position* of that point.

Sound waves are a particular form of so-called *elastic* waves. Whenever we produce a sudden deformation at a given place of a medium (e.g., when we hit a piano string with the hammer or when we suddenly displace air by starting the motion of the reed in a clarinet), elastic forces will cause the points close to the initial deformation to start moving. These points, in turn, will push or pull through elastic forces onto other neighboring points passing on to them the order to start moving, and so on. This "chain reaction" represents an elastic wave propagating away from the region of the initial perturbation. *What* propagates away with this wave is not matter but *energy*: that energy needed to put in motion each point reached by the wave. Sound waves of interest to music are elastic waves in which the points execute motions that are periodic. In its vibration each point of the medium always remains very, very close to the equilibrium position. A sound wave propagates with a well-defined speed away from the source in a straight line, until it is absorbed or reflected. The way sound waves propagate, are reflected, and absorbed determines the acoustical qualities of a room or concert hall.

We have mentioned above the concepts of force and energy. We must now specify their precise physical meaning. Everybody has an intuitive notion of *force:* the pull or the push we have to apply to change a body's shape, to set an object in motion, to hold a body in our hand, to slow a motion down, etc. But physics is not satisfied with intuitive concepts. We must give a clear definition of force, as well as the "recipe" of how to measure it. Both definition and recipe must be based on certain experiments whose results are condensed or summarized in the formulation of a physical law.

It is our daily experience that in order to change the form of a body we have to do something quite specific to it: we have to "apply a force." Deformation, i.e., a change in shape, is not the only possible effect of a force acting on a body. Indeed, it also is our daily experience that in order to alter the motion of a body we must apply a force. Quite generally it is found that the acceleration a of a body, representing the rate of change of its velocity, caused by a given force F is proportional to the latter. Or, conversely, the force is proportional to the acceleration produced: $F = ma$. This is called Newton's equation. The constant of proportionality m is the *mass* of the body. It represents its "inertia" or "resistance" to a change of motion. If more than one force is acting on a body, the resulting acceleration will be given by the *sum* of all forces. This sum may be zero; in that case the acting forces are in *equilibrium*.

The unit of force is defined as that force needed to accelerate a body of 1 kilogram at a rate of 1 m/sec². This unit of force is called the *Newton*. One Newton turns out to be equal to 0.225 pounds force. The pound is a unit of force (weight) most familiar to the public in English-speaking countries. Since the acceleration of gravity is 9.8 m/sec², the weight of a body of 1 kg mass turns out to be 9.8 Newton ($= 2.2$ lbs). We can measure a force by measuring the acceleration it imparts to a body of given mass or by equilibrating it (i.e., canceling its effect) with a known force, for instance, the tension in a calibrated spring.[1]

In many physical situations, a given force is applied or "spread" over an extended surface of the body. For instance, in a high-flying aircraft with a pressurized cabin, the air inside exerts a considerable outward-directed force, F, on

---

[1] "Calibrated" means that we have previously determined how much the spring stretches for a given force, e.g., a given weight.

each window (and on any other part of the hull), that is
proportional to the surface of the window, S. The relation
$p = F/S$ represents the *air pressure* inside the cabin. In general, we define the air pressure as the ratio between the
force acting on a surface S that separates the air from vacuum. If instead of vacuum we merely have a different pressure p' on the other side of the surface, the force F acting
on S will be given by

$$F = (p - p')S \qquad (3.1)$$

All this is very important for music. Sound waves in air are
air pressure oscillations. Thus, if in relation (3.1) S corresponds to the surface of the eardrum, p' is the (constant)
pressure in the middle ear, and p the oscillating pressure in
the meatus (Fig. 2.6), F will be the oscillating force acting
on the eardrum, responsible for its motion and that of the
bone chain in the middle ear.

The pressure is expressed in Newtons per square meter.
The normal atmospheric pressure at sea level is about
100,000 Newton/m². A more familiar unit is pound per
square inch (for instance, the overpressure in auto tires is
usually quoted in these units). Converting Newtons into lbs
and m² into sq in, we obtain 1 Newton/m² = 0.00015 lb/sq in.
Normal air pressure at sea level is thus about 15 lb/sq in.

We now turn to the concept of *energy*. Again, we do
have some intuitive idea about it—but our intuition may
easily fool us in this case. For instance, some people are
tempted to say that "it requires a lot of energy to hold a
heavy bag for a long time"—yet for the physicist no energy
is involved (except during the act of lifting the bag or put-
ting it down). For the physiologist, on the other hand, a
continuous flow of chemical energy to the muscles is necessary, to maintain a continued state of contraction of the
muscular fibers. To avoid confusion, it is necessary to introduce the concept of energy in a more precise, quantitative
way.

The concept of force alone does not suffice for the solu-
tion of practical problems in physics. For instance, we need
to know for how long, or over what distance, a given force
has been acting, if we want to determine, say, the final
speed acquired by a body accelerated by that force (even
the largest force may have only a small *end* effect, if the
duration or the path of its action was very short). As a
matter of fact, what really counts to determine a given
change of speed, say from a value 0 to v, is the product of
force times distance traveled in the direction of the force. If
we call x that distance, it can be shown mathematically,
based on Newton's equation, that $F \cdot x = \frac{1}{2}mv^2$. The product
$F \cdot x$ is called *work* and is counted positive if the displacement x is in the same direction as the force F. The product

$\frac{1}{2}mv^2$ is called the *kinetic energy* of the body of mass m. If $F \cdot x$ is positive we interpret the above relation by saying that the work of the force has increased the kinetic energy of the body, or, equivalently, that "work has been delivered to the system," increasing its kinetic energy from zero to $\frac{1}{2}mv^2$.

Work and kinetic energy are measured in Newton times meter. This unit is called *Joule*, after a British physicist and engineer. A body of 1 kg (unit of mass), moving at a speed of 1 m/sec thus has a kinetic energy of 0.5 Joule. If it moves with twice that velocity, its kinetic energy will be four times larger: 2 Joules. An average person (70 kg), running at a speed of 3 m/sec (6.75 miles/h) has a kinetic energy of 315 Joules; that of a 2,000 kg car cruising at 30 m/sec (67.5 miles/h) is 900,000 Joules.

Energy can appear in forms other than kinetic. Consider a body attached to a spring. We have to supply a given amount of work in order to compress the spring. If we do this very, very slowly, practically no kinetic energy will be involved. Rather, the supplied work will be converted into *potential energy*; in this case *elastic* potential energy of the body attached to a compressed spring. Releasing the spring, the body will be accelerated by the force of the expanding spring and the potential energy will be converted into kinetic energy. We may say that potential energy is the *energy of position* of a body, kinetic energy its *energy of motion*.

The sum of potential and kinetic energy of a body is called its total *mechanical energy* (there are many other forms of energy which we will not consider: thermal, chemical, electromagnetic, etc.). There are important cases in which the mechanical energy of a body remains constant. A "musically" important case is the previous example of a body attached to a spring oscillating back and forth under the action of the elastic force of the spring. It can be shown that the resulting vibration about the equilibrium position is *harmonic* (provided the amplitude remains small). When the body is released from a stretched position, its initial kinetic energy is zero. But it possesses an initial elastic potential energy which, as the oscillation starts, is converted into kinetic energy. Whenever the body is passing through its equilibrium position, the elastic potential energy is instantaneously zero while its kinetic energy is maximum. During the harmonic oscillation there is a back-and-forth conversion of potential energy into kinetic, and vice versa.

The total mechanical energy remains constant as long as no "dissipative" forces are in action. Friction causes a continuous decrease of total energy and hence a decrease of the amplitude of oscillation. The resulting motion is called a *damped oscillation*. It is extremely important in music. Indeed, many musical instruments involve damped oscillations; a vibrating piano string is a typical example. Other external forces may act in such a way as to gradually increase the mechanical energy. They can be used to compensate dissipative losses and thus maintain an oscillation at constant amplitude. A bowed violin string is a typical example: the forces that appear in the bowing mechanism feed energy into the vibrating string at a rate that is equal to the rate of energy loss through friction and acoustical radiation (Section 4.2).

Now we come to a last, but utmost important point concerning energy. Machines (and humans) deliver energy *at a given rate*. Any machine (or human) can perform an almost arbitrarily large amount of work—but it would take a very long time to do so! What really defines the "quality" or "power" of a machine is the *rate* at which it can deliver energy (i.e., perform work). This rate, if constant, is given by

$$P = \frac{\text{work done}}{\text{time employed}} = \frac{W}{(t_2 - t_1)} \qquad (3.2)$$

W is the work delivered between the times t_1 and t_2. P is called mechanical *power*. It is measured in units of Joule/sec, called *Watt* (another British engineer). If you are walking up a staircase, your body is delivering a power of about 600 Watt; the electrical energy consumed per second by an electric iron is about 1,000 Watt, and the maximum power delivered by a small car engine is 30 kilowatt (1 horsepower = 0.735 kW). A trombone playing fortissimo emits a total acoustical power of about 6 Watt.

The concept of power is most important for physics of music. Indeed, our ear is not interested at all in the total acoustical energy which reaches the eardrum—rather, it is sensitive to the *rate* at which this energy arrives, i.e., the acoustical *power*. This rate is what determines the sensation of *loudness*.

3.2 Propagation Speed, Wavelength, and Acoustical Power

After the excursion into the field of "pure" physics in the preceding section, we are in a better condition to understand the phenomenon of wave propagation. To that effect we make use of a *model* of the medium. We imagine the latter as made up of small bodies of given mass, linked to each other with compressed springs (representing the elastic forces). Initially, the spring forces are in equilibrium and all points are at rest. Figure 3.1 shows the situation when point P has been suddenly displaced an amount x_1 to the right.

Considering the forces shown in Fig. 3.1, we realize that both points Q and R—which initially are at rest at their respective equilibrium positions—are subject to a *resultant* force acting toward the right. In other words, according to Newton's equation they will be accelerated to the right and start a motion into the same direction in which P had originally been displaced. This point P, on the other hand, will

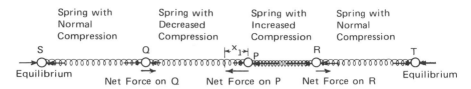

FIGURE 3.1. One-dimensional model of an elastic medium (springs in compression), in which point P has been displaced longitudinally.

be on its way back to its equilibrium position, accelerated by a resultant force that acts on it toward the left (Fig. 3.1). A short time later, when points Q and R are on their way toward the right, the compression of the spring between R and T starts increasing, whereas that of the spring between Q and S decreases. It is easy to see that both points S and T will start being subjected to a net force directed to the right that will cause them to start moving to the right, while Q and R may be on their way back to the left. This process goes on and on, from point to point—representing a wave propagating away from P toward both sides. The wave "front" is nothing but an "order" that goes from point to point telling it: "start moving to your right." The "order" is given by the compressed springs (their elastic forces). At no time is there any net transport of matter involved. We call this case a *longitudinal wave*, because the displacements of the points are directed parallel to the direction of propagation of the waves. In the real case of a sound wave propagating through air, the concerted action of the spring forces acting on points $P, Q, R, \ldots$ roughly corresponds to the air pressure; variations of these forces (e.g., variations of the distances between points) correspond to the *air pressure variations* of the sound wave.

The one-dimensional model of Fig. 3.1 also shows how energy transport is involved in an elastic wave. In first place, we have to provide work from "outside" to produce the initial displacement x_1 of point P because we have to modify the lengths of the two springs PQ, PR. In other words, we need an energy source. In this case the initial energy is converted into the form of potential (positional) energy of point P. Then, as time goes on, points to the right and to the left of P start moving, and the lengths of their springs change. All these processes involve energy, both kinetic (motion of the points) and potential (compression or expansion of springs). The energy initially given to P is transferred from point to point of the medium, as the wave propagates: we have a *flow* or transport of energy away from the source.

Let us now turn to the case in which the springs in the model are in tension (expanded) instead of being compressed, with neighboring points *pulling* on each other. Physically, this corresponds to a tense violin string. For longitudinal displacements (in the direction along the springs) we obtain a qualitatively similar picture for wave propagation as before, only that all forces shown in Fig. 3.1 are now reversed. But in addition we have an entirely new possibility that does not exist for the case of compressed springs: we may displace point P *perpendicularly* to the x direction (Fig. 3.2) and obtain a different type of wave. Since all spring forces now pull on the points, according to Fig. 3.2, the resultant force F_P will accelerate P down to its equilibrium position O. Points Q and R in turn would be subject to net forces that would accelerate them upward, in a direction essentially perpendicular to x. This represents a *transverse* elastic wave, propagating to the right and to the left of P. In a transverse wave, the displacements of the points are perpendicular to the direction of propagation. In a medium under tension, like a violin string, *two* modes of elastic wave propagation may occur simultaneously: transverse and longitudinal.

We now turn to the expression for the speed of propagation of transverse waves. It can be shown by applying Newton's law to the individual points of the unidimensional model of Fig. 3.2 that, for a string under a tension T (in Newton), the velocity V_T of transverse elastic waves is given by:

$$V_T = \sqrt{\frac{T}{d}} \quad \text{(m/sec)} \tag{3.3}$$

d is the "linear density" of the medium, i.e., *mass per unit length* (in kg/m). Notice that the more tense a string is, the faster the transverse waves will travel. On the other hand, the more dense it is, the slower the waves will propagate.

A physically equivalent relation exists for the propagation speed of longitudinal waves in a medium of density δ (in kg/m³) and where the pressure is p in Newton/m²:

$$V_L = \sqrt{\frac{p}{\delta}} \quad \text{(m/sec)} \tag{3.4}$$

FIGURE 3.2. One-dimensional model of an elastic medium (springs in expansion), in which point P has been displaced transversally.

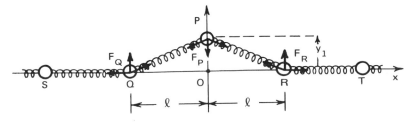

For an ideal gas, however, the ratio p/δ turns out to be proportional to the "absolute" temperature t_A, defined in terms of the centigrade or Fahrenheit temperatures t_C and t_F by the simple transformation:

$$t_A = 273 + t_C = 273 + \tfrac{5}{9}(t_F - 32) \qquad \text{(degrees Kelvin)} \quad (3.5)$$

Notice that at the freezing point ($t_C = 0°C$, $t_F = 32°F$), the absolute temperature is $t_A = 273°$. Although ordinary air is not a 100% "ideal gas," it behaves approximately so, and the velocity of sound waves may be expressed as

$$V_L = 20.1 \ \sqrt{t_A} \qquad \text{(m/sec)} \qquad\qquad (3.6)$$

This turns out to be 331.5 m/sec ($= 1087$ feet/sec) at 0°C (32°F) and 334 m/sec ($= 1130$ feet/sec) at 21°C (70°F). The numerical factor in (3.6) is for air only. In general, its value depends on the *composition* of the medium through which the sound propagates. For pure hydrogen, for instance, it is equal to 74.0. Sound waves thus travel almost four times as fast in hydrogen as in air. This leads to funny acoustical effects if a person speaks or sings after having inhaled hydrogen (NO SMOKING!)

Sound travels fast, but not infinitely fast. This, for instance, leads to small but noticeable *arrival time differences* between sound waves from different instruments in a large orchestra and may cause serious problems of rhythmic synchronization. A pianist who for the first time plays on a very large organ, in which the console is far away from the pipes, initially may become quite confused by the late arrival of the sound, out of synchronism with his fingers. *Reverberation* in a hall is based on superposition of delayed sound waves that have suffered multiple reflections on the walls (Section 4.7).

Let us now consider a very long string in which the initial point is set in vibration with simple harmonic motion and continues to vibrate indefinitely compelled by some external force. After a while, all points of the string are found to vibrate with the same simple harmonic motion. If at a given instant of time the initial point is, say, at its maximum displacement, its neighbors are not yet quite there, or just had been there, and so on. Figure 3.3 shows the transverse displacements of all points of the string *at a given time*. This curve is a "snapshot" of the shape of the string during the passage of a sinusoidal transverse wave. The graph in Fig. 3.3 should not be confused with the curve shown in Fig. 2.4, which represents *the time history of only one given point*. The latter indicates a vibration pattern in time, the former a wave pattern in space. The shortest distance between any two points of the string that are vibrating in a parallel way (vibrating "in phase," i.e., having identical displacements y at all times), is called *wavelength*. It is usually designated

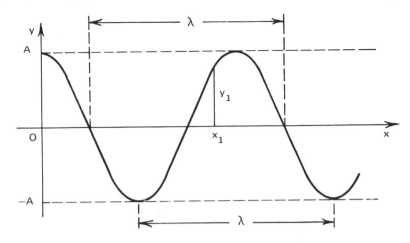

FIGURE 3.3

with the Greek letter λ. Alternatively the wavelength can be defined as the minimum space interval after which the spatial wave pattern repeats. Compare this with the definition of the period, which represents the minimum *time* interval after which the vibration pattern of *one given point* repeats (Fig. 2.3b).

As time goes on, the snapshot curve seems to move with the speed of the wave to the right (Fig. 3.4)—yet each point of the string only moves up and down (for instance, consider point x_1 in Fig. 3.4). *What* moves to the right is the configuration, i.e., the actual shape of the string, but not the string itself. In other words, what moves to the right is a quality, for instance, the quality of "being at the maximum displacement" (e.g., points P, Q, R in Fig. 3.4), or the quality of "just passing through equilibrium" (points S,T,U). And, of course, what also moves to the right is *energy*, the potential and kinetic energy involved in the up-and-down oscillation of the points of the string.

There is an important relationship between the speed V of a sinusoidal wave, its wavelength λ, and the frequency f of oscillation of the individual points. Considering Fig. 3.3, we realize that the wave will have moved exactly one wavelength λ during the time it takes the initial point (or any other) to make one complete oscillation, i.e., during one period τ. We therefore can write for the speed of the wave:

$$V = \frac{\text{distance traveled}}{\text{time employed}} = \frac{\lambda}{\tau}$$

Since the inverse of the period is equal to the frequency f [relation (2.1)], we can also write

$$V = \lambda \cdot f \tag{3.7}$$

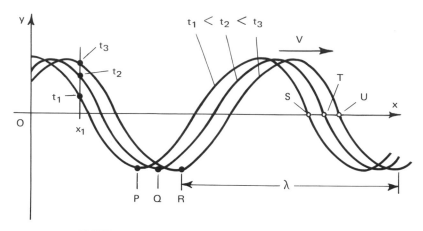

FIGURE 3.4

This relation provides the quantitative link between the "space representation" of Fig. 3.3 and the "time representation" of Fig. 2.4. Relation (3.7) enables us to express the wavelength of a transverse wave in a string in terms of the frequency of the oscillation of the individual points and the propagation velocity (3.3):

$$\lambda = \frac{1}{f} \sqrt{\frac{T}{d}} \tag{3.8}$$

In the case of *longitudinal waves* such as a sound wave the points vibrate in a direction *parallel* to the direction of propagation, and it is not so easy to picture their real position in visual form. For this reason, sound waves are more conveniently represented as pressure oscillations. Figure 3.5 (bottom row) shows the displacements of the points of a unidimensional model of the medium, when a longitudinal wave is passing through. Notice that points show their maximum accumulation (i.e., maximum pressure) and maximum rarefaction (i.e., minimum pressure) at the places where their displacement is zero (points *P, Q,* respectively). On the other hand, at places where the displacements are maximum, the pressure variations are zero. This means that the pressure variations of a sound wave are 90° out of phase with the oscillation of the points: the maximum pressure variations (either increase or decrease) occur at places where the displacements of the points are zero; conversely, maximum displacements of the points occur at places where the pressure variations are zero.

A sinusoidal sound wave is one in which the pressure at each point oscillates harmonically about the "normal" (undisturbed) value (Fig. 3.6). At a point like *A*, all points of the medium have come closest to each other (maximum

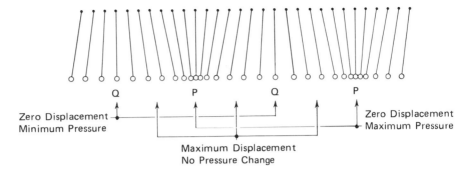

Q   P   Q   P

Zero Displacement
Minimum Pressure

Zero Displacement
Maximum Pressure

Maximum Displacement
No Pressure Change

FIGURE 3.5. Longitudinal wave in a one-dimensional medium.
To show the actual displacements, each point is depicted
as the bob of a pendulum.

pressure increase, points *P* in Fig. 3.5); at a point like *B*,
they have moved away from each other (maximum pressure
decrease, points *Q* in Fig. 3.5). The *average pressure variation*, Δ*p*, is equal to the pressure variation amplitude divided by $\sqrt{2}$ ($= 1.41$). Taking into account relations (3.6)
and (3.7) we obtain for the wavelength of sinusoidal sound
waves in air:

$$\lambda = \frac{20.1}{f} \sqrt{t_A} \qquad \text{(in meters)} \qquad (3.9)$$

t_A is the "absolute" temperature given by (3.5). Typical
values of wavelengths at normal temperature are shown in
Fig. 3.7.

Elastic waves can be transmitted from one medium to
another—for instance, from air into water, from air into a
wall and then again into air, from a string to a wooden
plate, and from there to the surrounding air. The nature of
the wave may change in each transition (e.g., the transition
from a transverse wave in the string and plate to a longi-

FIGURE 3.6

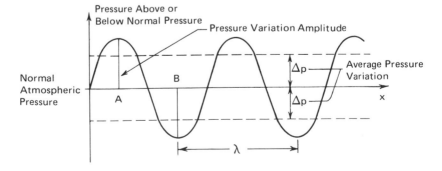

Pressure Above or
Below Normal Pressure
Pressure Variation Amplitude

Normal
Atmospheric
Pressure

Average Pressure
Variation

Δp

Δp

A

B

λ

x

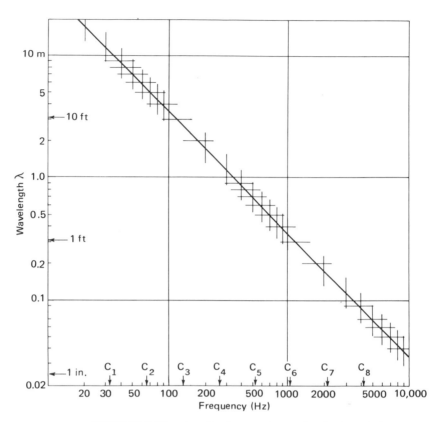

FIGURE 3.7. Wavelength of a sound wave in air at normal temperature, as a function of frequency (logarithmic scale).

tudinal sound wave in the air). However, in each transition, the *frequency remains invariant*. The wavelength, on the other hand, will change according to relation (3.7): $\lambda = V/f$. In this relation, V changes from medium to medium, while f is dictated exclusively by the initial vibration (source).

When an elastic wave hits the boundary between two media, part of it is *reflected* back to the original medium. Some boundaries are almost perfect reflectors (e.g., smooth cement walls, for sound waves; the fixed end points of a tense string for transverse waves). This phenomenon is governed by the fact that on the reflecting boundary the points of the medium are compelled to remain at rest, therefore upsetting the balance of elastic forces that "command" the wave propagation. In a reflection, the frequency remains unchanged, while the direction of propagation is reversed for a perpendicular incidence (or, in general, directed with a reflection angle equal to the incidence angle). Also the amplitude would remain the same if there were no absorption.

We finally consider the *energy flow* associated with a sound wave. We define it as the amount of total mechanical energy (potential and kinetic, associated with the elastic oscillations of the points of the medium) that is transferred during each second through a surface of unit area (1 m²) perpendicular to the direction of propagation (Fig. 3.8). This energy flow is expressed in Joule per m² and sec, or, taking into account the definition and the units of power (3.2), in Watt/m². It is more commonly called the *intensity* of the wave, and designated with the letter I. It can be shown that there is a relation between the intensity of a sinusoidal sound wave and the value of the *average pressure oscillation* associated with the wave (see Fig. 3.6), which we denote with Δp (equal to the pressure variation amplitude divided by $\sqrt{2}$):

$$I = \frac{(\Delta p)2}{V\delta}$$

In this relation, V is the speed of the sound wave (3.6) and δ is the air density. For normal conditions of temperature and pressure, we have the following numerical relationship:

$$I = 0.00234 \times (\Delta p)^2 \quad \text{(Watt/m}^2) \tag{3.10}$$

Δp must be expressed in Newton/m². As we shall see in Section 3.4, the faintest pure sound that can be heard at a frequency of 1,000 Hz has an intensity of only 10^{-12} Watt/m². According to relation (3.10) this represents an average pressure variation of only 2.0×10^{-5} Newton/m², i.e., only 2.0×10^{-10} that of the normal atmospheric pressure! This gives an idea of how sensitive the ear is.

A given sound source (a musical instrument or a loudspeaker) emits sound waves into all directions. In general, the amount of energy emitted per second depends on the particular direction considered. Let I_1 be the intensity of the wave at point A_1 propagating along the direction shown in Fig. 3.9. This means that an amount of energy $I_1 a_1$ flows through the surface a_1 during each second. If we assume that no energy is lost on its way, this same amount of energy

FIGURE 3.8

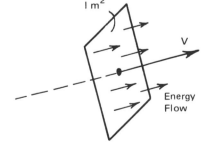

FIGURE 3.9

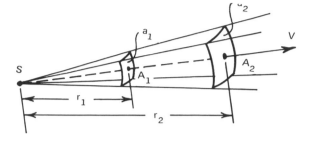

will flow each second through surface a_2 at point A_2. There-fore

$$I_1 a_1 = I_2 a_2$$

Since the areas of the surfaces a_1 and a_2 are proportional to the squares of their respective distances r_1 and r_2 to the source, the intensity of a sound wave varies inversely pro-portional to the square of the distance to the source:

$$\frac{I_1}{I_2} = \left(\frac{r_2}{r_1}\right)^2 \tag{3.11}$$

This law is no longer true if we take into account sound reflections and absorption.

If we imagine the whole sound source encircled by a spherical surface, the total amount of energy flowing each second through that surface is called the *acoustic power out-put* of the source. It represents the rate at which the source emits energy into *all* directions in form of sound waves. Its value is given in Watts (Joule/sec). Typical instruments radi-ate between 0.01 Watt (clarinet) up to 6.4 Watts (trombone playing fortissimo).

**3.3
Superposition
of Waves;
Standing Waves**

In absence of reflecting walls, sound waves travel in straight lines away from the source. As shown in the previous sec-tion, their intensity decreases rapidly, proportional to $1/r^2$, where r is the distance to the source. If we have more than one source, the waves emitted by each one will propagate individually as if no other wave would exist, and the resultant effect at one given point of the medium (for instance in the auditory canal) will be a pressure oscillation that is simply given by the algebraic sum of the pressure oscillations of the individual waves.[2] In other words, *sound waves superpose linearly.*

Let us consider the superposition of two pure sound

---

[2] Note carefully that what is added here are pressure *variations*, and not absolute values of the pressure!

waves of frequencies f_1 and f_2 and, according to relation (3.7), of wavelengths $\lambda_1 = V/f_1$, $\lambda_2 = V/f_2$, traveling in the *same* direction. In order to obtain a "snapshot" of the resulting pressure variations, we just have to add the values of the individual pressure variations, as caused by each wave separately at each point x along the direction of propagation. Since the velocity of sound waves does not depend on frequency (nor on the vibration pattern as a whole), all points of the medium will repeat exactly the same complex vibration pattern—only subject to a different timing. The energy flow—i.e., the intensity of the superposition of two (or more) waves traveling in the same direction with random phases—is simply the sum of the energy flow contributions from the individual components:

$$I = I_1 + I_2 + I_3 + \cdots \tag{3.12}$$

A particularly important case is that of two sinusoidal waves of the same frequency and the same amplitude *traveling in opposite directions*. This, for instance, happens when a sinusoidal wave is reflected at a given point (without absorption) and then travels back, superposing itself with the incoming wave. Let us first consider transverse waves in a string (Fig. 3.10). Adding the contributions from each component, we obtain another sinusoidal wave of the same frequency but of different amplitude. The striking fact, however, is that this resultant wave *does not propagate at all!* It remains anchored at certain points N_1, N_2, N_3 . . . . called *nodes*, which do not vibrate. Points between nodes vibrate with different amplitudes, depending on their position. In particular, points A_1, A_2, A_3 . . . (midway between nodes), called *antinodes*, vibrate with a maximum amplitude (twice that of each component wave). Figure 3.11 shows the successive shapes of a string when two sinusoidal waves of the same amplitude travel in opposite directions. This is called a *standing wave*. Points oscillate, but there is no evidence of propagation. The wave profile changes in amplitude, but does not move neither to the right nor to the left. At one time (t_1) the string shows a maximum deformation; at another (t_5) it has no deformation at all. As we shall see in the next chapter, standing waves play a key role in music, especially in the sound generation mechanisms of musical instruments.

In a standing wave *there is no net propagation of energy*, either. The whole string almost acts as one elastic vibrating spring: at one given time (e.g., t_5 in Fig. 3.11), all points are passing through their equilibrium position, and the energy of the whole string is in kinetic form (energy of motion). At another instant (e.g., t_1 in Fig. 3.11), all points are at their maximum displacement and the energy is all potential. In other words, in a standing wave all points oscillate in phase.

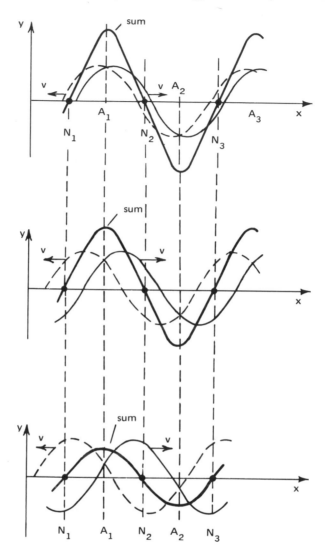

FIGURE 3.10

Note carefully that this does not happen with a propagating wave: in Fig. 3.3, for instance, at one given instant of time there are points which have maximum displacement (only potential energy) as well as points with zero displacement (only kinetic energy), or points in any intermediate situation (both forms of energy). Furthermore, in a propagating wave all points have the same amplitude; what varies are the times at which a maximum displacement is attained (the points are out of phase).

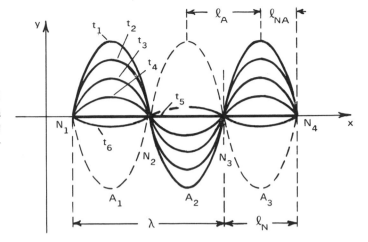

FIGURE 3.11

Successive shapes of a string in a standing wave oscillation.

A careful inspection of Fig. 3.11 reveals that the distance l_N between two neighboring nodes, N_1, N_2, or the distance l_A between two antinodes A_1, A_2 is exactly one half of a wavelength λ:

$$l_N = l_A = \lambda/2 \tag{3.13}$$

On the other hand, the distance l_{NA} between a node N_1 and an antinode A_1 is a quarter wavelength:

$$l_{NA} = \lambda/4 \tag{3.14}$$

Standing waves can also be longitudinal. They arise when two sound waves of the same frequency and pressure variation amplitude travel in opposite directions. This happens, for instance, when a sound wave travels along a pipe and is reflected at the other end; standing waves also arise from reflections on the walls in rooms and halls. They have the same properties as transverse standing waves, and the discussion given above applies here, too. There is, however, an important additional remark to be made. As pointed out in the previous section, sound waves are most conveniently described by pressure oscillations. We showed there that points with maximum pressure variation have zero longitudinal displacement (Fig. 3.5), whereas places with zero pressure variations correspond to points with maximum displacement. We can translate this to the case of a standing sound wave: *pressure nodes* (i.e., points whose pressure variations are permanently zero) *are vibration antinodes* (points which oscillate with maximum amplitude), whereas *pressure antinodes* (points at which the pressure oscillates with maximum amplitude) *are vibration nodes* (points which remain permanently at rest).

**3.4
Intensity, Sound
Intensity Level,
and Loudness**

In Section 2.3 we stated that, for a pure sound, the amplitude of the eardrum oscillations leads to the sensation of loudness. This amplitude is directly related to the average pressure variation, Δp, of the incoming sound wave and, hence, to the acoustical energy flow or intensity I reaching the ear [relation (3.10)]. We start here by investigating the range of intensities I of pure sound waves to which the ear is sensitive. There are two limits of sensitivity to a tone of given frequency: (1) a lower limit or *threshold of hearing* representing the minimum just audible intensity; (2) an *upper limit of hearing* beyond which physiological pain is evoked, eventually leading to physical damage of the hearing mechanism. One finds that these two limits vary from individual to individual and depend on the particular frequency under consideration. In general, for a tone of about 1,000 Hz (a pitch between the notes B_5 and C_6), the interval between limits is largest. The enormous range of intensities encompassed between the two limits of hearing is startling. Indeed, for a tone of 1,000 Hz, it is found that the average threshold intensity lies near 10^{-12} Watt/m², whereas the limit of pain lies at about 1 Watt/m². This represents a ratio of intensities of one trillion to one, to which the ear is sensitive! Table 3.1 shows the relations between sound intensity and musical loudness sensation, for a 1,000 Hz tone.[3] At 1,000 Hz, the intensity range of *musical* interest extends from about 10^{-9}

Table 3.1

Intensity (Watt/m²)	Loudness[3]
1	Limit of pain
10^{-3}	*fff*
10^{-4}	*ff*
10^{-5}	*f*
10^{-6}	*mf*
10^{-7}	*p*
10^{-8}	*pp*
10^{-9}	*ppp*
10^{-12}	Threshold of hearing

[3] It is rather arbitrary to consider the music notation as representing an "absolute" measure of loudness. Musicians for instance will argue that we are perfectly able to perceive fortissimos and pianissimos in music played on a radio with its "volume" control turned down to a whisper. What happens in such a case is that we use cues *other* than intensity to make subjective judgments of "relative" loudness. On the other hand, systematic experiments (Patterson 1974) have revealed that the interpretation of the musical loudness notation in an actual dynamic context is highly dependent on the instrument and on the pitch range covered.

to 10^{-2} Watt/m^2. This still represents a variation of a factor of ten millions!

Because of this tremendous range, the unit of Watt/m^2 is impractical. There is another reason why it is impractical. The just noticeable difference (jnd) of a given stimulus is usually a good physical "gauge" to take into account when it comes to choose an appropriate unit for the corresponding physical magnitude. Experiments show that the *jnd in tone intensity* is roughly proportional to the intensity of the tone. This proportionality thus suggests that the appropriate "unit" should gradually increase, as the intensity of the tone we want to describe increases. This, of course, would lead to an awful complication unless we introduce a different magnitude which is an appropriate function of the intensity I. This new magnitude should accomplish three simultaneous objectives: (1) a "compression" of the whole audible intensity scale into a much smaller range of values; (2) the use of relative values (for instance, relative to the threshold of hearing) rather than absolute ones; and (3) the introduction of a more convenient unit whose value closely represents the minimum perceptible change of sound intensity.

The introduction of the new quantity is done in the following way. Note in Table 3.1 that what seems more characteristically related to the loudness effect is the *exponent* to which the number 10 is raised when we quote the value of the sound intensity (left column); -12 for the threshold of hearing, -9 for a *ppp* sound, -7 for piano, -5 for forte, -3 for forte-fortissimo, and 0 for limit of pain ($10^0 = 1$). This strongly suggests that we should use what in mathematics is called a logarithmic function to represent intensity.

The *decimal logarithm* of a given number is the exponent to which 10 has to be raised in order to yield that number. For instance, the logarithm of 100 is 2 because $10^2 = 100$; the logarithm of 10,000 is 4 because $10^4 = 10,000$; the logarithm of 1 is zero, because $10^0 = 1$; and the logarithm of 0.000001 is -6 because $10^{-6} = 0.000001$. These relations are written symbolically: log $100 = 2$; log $10,000 = 4$; log $1 = 0$; log $0.000001 = -6$. For any number intermediate between integer powers of ten, the logarithm can be found through the use of tables.

An important property is that the logarithm of the *product* of two numbers is the *sum* of the logarithms of the individual numbers. For instance, the logarithm of the number 10^4 times 10^3 is 4 *plus* 3 (i.e., 7), because $10^4 \times 10^3 = 10^{4+3} = 10^7$. In general, for any two numbers a and b, we have the relation log $(a \times b) = $ log $a +$ log b. For the logarithm of a division a/b, we have, instead, log $(a/b) = $ log $a -$ log b.

Decimal logarithms can be used to define a more appropriate magnitude to describe sound intensity. First of all, we adopt the hearing threshold (at 1,000 Hz) of 10^{-12} Watt/m² as our *reference* intensity I_0, Then we introduce the quantity

$$IL = 10 \times \log \frac{I}{I_0} \qquad (3.15)$$

This is called the *sound intensity level*. The unit of IL is called the *decibel*, denoted "db." For the hearing threshold, $I/I_0 = 1$ and $IL = 0$ db. For the upper limit of hearing $I/I_0 = 10^{12}$ and $IL = 10 \times \log 10^{12} = 120$ db. A typical "forte" sound (Table 3.1) has a sound intensity level of 70 db; *ppp* corresponds to 30 db.

It is important to note that when a quantity is expressed in decibels, a *relative* measure is given, with respect to some reference value (e.g., the hearing threshold in the definition of IL). Whenever the intensity I is multiplied by a factor of ten, one just *adds* ten decibels to the value of IL; when the intensity is multiplied by 100, one must add 20 db, etc. Likewise, when the intensity is divided by a factor of 100, one must *subtract* 20 db from IL. Table 3.2 gives some useful relationships.

We may use relation (3.10) to express the intensity in terms of the much more easily measurable average pressure variation, Δp. We find that the minimum threshold I_0 at 1,000 Hz roughly corresponds to an average pressure variation $\Delta p_0 = 2 \times 10^{-5}$ Newton/m². Since according to relation (3.10) I is proportional to the *square* of Δp, we have

$$\log \frac{I}{I_0} = \log \left(\frac{\Delta p}{\Delta p_0} \right)^2 = 2 \times \log \frac{\Delta p}{\Delta p_0}$$

Hence, one may introduce the quantity

$$SPL = 20 \times \log \frac{\Delta p}{\Delta p_0} \qquad (3.16)$$

called the *sound pressure level* (SPL). For a traveling wave, the numerical values of (3.15) and (3.16) are identical, and IL and SPL represent one and the same thing. For *standing* waves, however, there is no energy flow at all (Section 3.3) and the intensity I as used in (3.15) cannot be defined; hence

Table 3.2

Change in IL	What happens to the intensity
add (subtract)　1 db	Multiply (divide) by 1.26
+(−)　3 db	×(+) 2
+(−) 10 db	×(+) 10
+(−) 20 db	×(+) 100
+(−) 60 db	×(+) 1,000,000

IL loses its meaning. Yet the concept of average pressure variation Δp at a given point in space (e.g., at the entrance to the auditory canal) still remains meaningful and so does the sound pressure level. This is why relation (3.16) is more frequently used than (3.15). Note carefully that the definitions of IL and SPL do not involve at all the frequency of the sound wave. Although we did make reference to a 1,000 Hz tone, nothing prevents us from defining IL and SPL through relations (3.15) and (3.16), respectively, for any arbitrary frequency. What does depend on frequency, and very strongly, are the subjective limits of hearing (e.g., I_0 and p_0) and, in general, the *subjective* sensation of loudness, as we shall see further below.

Funny things seem to happen with the sound intensity level, or sound pressure level, when we superpose two sounds of the same frequency (and phase). Just consider Table 3.2: adding two tones of the same intensity, which according to (3.12) means doubling the intensity, adds a mere 3 db to the sound level of the original sound, whatever the actual value of that IL might have been. Superposing *ten* equal tones (in phase) only increases by 10 db the resulting IL. To raise the IL of a given tone by 1 db, we must multiply its intensity by 1.26, meaning that we must add a tone whose intensity is 0.26 (about ¼) that of the original tone.

The minimum change in SPL required to give rise to a detectable change in the loudness sensation (*jnd* in sound level) is roughly constant and of the order of 0.2–0.4 db in the musically relevant range of pitch and loudness. The unit of IL or SPL, the decibel, is thus indeed of "reasonable size"— close to the *jnd*.

There is an alternative way of looking at the *jnd* of intensity or sound level. Instead of asking how much the intensity of *one* given tone must be changed to give a just noticeable effect, we may pose the totally equivalent question: what is the minimum intensity I_2 that a *second* tone of the same frequency and phase must have, to be noticed in presence of the first one (whose intensity I_1 is kept constant)? That minimum intensity I_2 is called the *threshold of masking*. The original tone of constant intensity I_1 is called the "masking tone," the additional tone is the "masked tone." Masking plays a key role in music. In this paragraph we only mention the masking of tones of frequency (and phase) identical to that of the masking tone; further below we shall discuss masking at different frequencies. The relation between the masking level ML (IL of the masked tone at threshold) and the *jnd* of sound level can be found using relation (3.15) and taking into account the properties of logarithms. For instance, for a *jnd* of 0.2 db one obtains a masking level ML that is 13 db less than the IL of the masking tone; a *jnd* of 0.4 db corresponds to a ML that is 10 db below.

So far we have been dealing with the physical quantities *IL* and *SPL*. Now we must examine the psychological magnitude *loudness,* associated with a given *SPL*. In Sections 1.4 and 2.3 we mentioned the ability of individuals to establish an order for the "strength" of two sensations of the same kind, pointing out that complications arise when absolute quantitative comparisons are to be made. In the case of loudness, judgments of whether two pure tones sound equally loud show fairly low dispersion among different individuals. But judgments as to "how much" louder one tone is than another require previous conditioning or training and yield results that fluctuate greatly from individual to individual.

Tones of the same *SPL* but with different frequency in general are judged as having different loudness. The *SPL* is thus not a good measure of loudness, if we intercompare tones of different frequency. Experiments have been performed to establish *curves of equal loudness,* taking the *SPL* at 1,000 Hz as a reference quantity.[4] These are shown in Fig. 3.12 (Fletcher and Munson 1933). Starting from the vertical axis centered at 1,000 Hz toward both sides of lower and higher frequencies, respectively, are drawn curves that correspond to *SPL*'s of tones that are judged "equally loud" as the reference tone of 1,000 Hz. Note, for instance, that while a *SPL* of 50 db (intensity of 10^{-7} Watt/m²) at 1,000 Hz is considered "piano," the same *SPL* is barely audible at 60 Hz. In other words, to produce a given loudness sensation at low frequencies, say a "forte" sound, a much higher intensity (energy flow) is needed than at 1,000 Hz. This is why bass tones seem to "fade away" much before the trebles, when we gradually move away from a fixed sound source. Or why we have to pay so much more for a hi-fi set, particularly the speakers, if we want well-balanced basses.

The lowest curve in Fig. 3.12 represents the threshold of hearing for different frequencies. Again, it shows how the sensitivity of the ear decreases considerably toward low (and also toward very high) frequencies. Maximum sensitivity is achieved at about 3,000 Hz. The shape of this threshold curve is influenced by the acoustical properties of the auditory canal (meatus) and by the mechanical properties of the bone chain in the middle ear. Finally, it should be emphasized that the curves in Fig. 3.12 are valid only for *single, continuously sounding pure tones.* We shall discuss further below what happens to the loudness sensation if a tone is of short duration (less than a second). Recent studies (Molino 1973) show that equal loudness contours appear to depend

---

[4] Obtained through "loudness matching" experiments conducted in a way quite similar to pitch matching experiments.

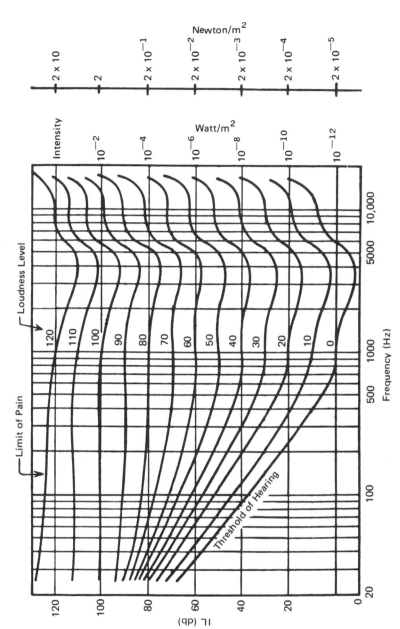

FIGURE 3.12. Curves of equal loudness (Fletcher and Munson 1933) in a sound intensity level (IL) and frequency diagram. The corresponding scales of sound intensity and average pressure variation are also shown.

Reprinted by permission from the Journal of the Acoustical Society of America.

on the frequency of the reference tone (which was 1,000 Hz in Fig. 3.12).

Now comes a sometimes confusing issue. A new quantity is introduced, called the *loudness level*, which we designate *LL*. It is defined in the following way. The *LL* of a tone of frequency *f* is given by the *SPL* of a tone of 1,000 Hz that is judged to be equally loud. This means that the curves of Fig. 3.12 are curves of constant loudness level. The unit of *LL* is called the *phon*. Figure 3.12 can be used to find the *LL* of a tone of given *SPL*, at any frequency *f*. For instance, consider a tone of 70 db *SPL* ($I = 10^{-5}$ Watt/m^2) at 80 Hz. We see that the curve that passes through that point intersects the 1,000 Hz line at 50 db. The *LL* of that tone is thus equal to 50 phons. In general, the numbers shown along the 1,000 Hz line represent the *LL* in phon of the corresponding constant loudness curves.

Note very carefully that *LL* still is a *physical* magnitude, rather than a psychophysical one (in spite of the name). It represents those intensities or *SPL's* that sound equally loud, but it does not pretend to represent loudness in an absolute manner: a tone whose *LL* is twice as large simply does not sound twice as loud! Many studies have been made to determine a subjective scale of loudness. Figure 3.13 (solid line) is the result (Stevens 1955), relating the "subjective loudness" *L* with the *loudness level LL*, in the range of musical interest. The quantity *L* describing subjective loudness is ex-

FIGURE 3.13. Solid line: experimental relation between the psychological magnitude loudness and the physical magnitude loudness level (after Stevens 1955). Broken line: power-law relationship (3.17) (Stevens 1970).

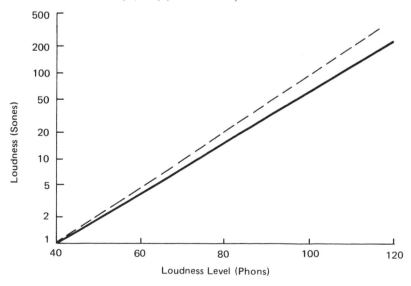

pressed in units called *sones*. Notice that the relationship is not linear (the loudness scale in Fig. 3.13 is what one calls a logarithmic scale). It is such that increasing LL by 10 phons, the loudness L is merely *doubled*. This, for instance, means that *ten* instruments playing a given note at the same LL are judged to sound only *twice* as loud as one of the instruments playing alone!

It has been shown that the relation between L and the wave intensity I, or the average pressure variation Δp, can be described approximately by the simple function (Stevens 1970):

$$L = C_1 \sqrt[3]{I} = C_2 \sqrt[3]{(\Delta p)^2} \tag{3.17}$$

where C_1 and C_2 are parameters that depend on frequency. This yields the broken line shown in Fig. 3.13, which lies well within the statistical fluctuation of the actual measurements (not shown). Notice that the logarithmic relationship [as in (3.15) and (3.16)] is all but gone. Yet an appreciable "compression" of the subjective loudness scale still remains: in order to vary L between 1 and 200, the intensity I must change by a factor of 8 million.

When we superpose two or more tones of the same frequency (and randomly mixed phases), the resulting tone has an intensity (energy flow) which is the sum of the intensities of the component tones: $I = I_1 + I_2 + I_3 + \cdots$ (3.12). Since in this case the individual tones cannot be discriminated from each other, this total intensity determines the resulting loudness through relation (3.17). L obviously will not be equal to the sum of the loudnesses of the individual tones. Different situations arise when the component waves have *different frequencies*. We may distinguish three regimes: (1) If the frequencies of the component tones fall all *within the critical band* of the center frequency (Section 2.4), the resulting loudness still is directly related to the total intensity (energy flow), sum of the individual intensities:

$$L = C_1 \sqrt[3]{(I_1 + I_2 + I_3 + \cdots\cdots)} \tag{3.18}$$

This property actually leads to a far more precise determination (Zwicker, Flottorp, and Stevens 1957) of the critical band than that given in Section 2.4. (2) When the frequency spread of the multitone stimulus *exceeds the critical band*, the resulting subjective loudness is *greater* than that obtained by simple intensity summation (3.18), increasing with increasing frequency difference and tending toward a value that is given by the sum of individual loudnesses:

$$L = C_1 \sqrt[3]{I_1} + C_1 \sqrt[3]{I_2} + C_1 \sqrt[3]{I_3} + \cdots \tag{3.19}$$

Masking effects must be taken into account if the individual loudnesses L_1, L_2, etc., differ considerably among each other.

The limit of loudness integration (3.19) is never achieved in practice. (3) When the frequency difference between individual tones is large, the situation becomes complicated. First of all, difficulties arise with the concept of "total loudness." People tend to focus on only *one* of the component tones (e.g., the loudest, or that of highest pitch) and assign the sensation of total loudness to just that single component:[5]

$$L = \left. \begin{matrix} \text{maximum} \\ \text{or highest} \end{matrix} \right\} \text{ of } (L_1, L_2, \ldots) \tag{3.20}$$

All this has very important consequences for music. For instance, two organ pipes of the same kind and the same pitch sound only 1.3 times louder than one pipe alone. When their pitch is a semitone or a whole tone apart their loudness still will be roughly 1.3 times that of one single pipe (a semitone or whole tone falls within the critical band, Fig. 2.13). But two tones that are a major third apart will sound louder than the previous combination. These facts have been well known for centuries to organ builders and composers as well. Since there is no possibility for a manual loudness control of individual organ tones, as in string instruments or woodwinds, loudness on the organ can be altered only by changing the number of simultaneously sounding pipes. But since according to the previous paragraph, loudness summation is more effective when the component tones differ in frequency appreciably, stops sounding one (4′ stop), two (2′), or more octaves above [and below (16′)] the written note are mainly used for that purpose.[6] On the other hand, without adding or subtracting stops, loudness may also be controlled through the number of notes simultaneously played. Each new voice entering in a fugue increases the subjective loudness of the piece, and each additional note in a chord accomplishes the same purpose. Some organists play the final chords of a Bach fugue pulling additional stops. This is absurd—Bach himself programmed the desired loudness increase by simply writing in more notes than the number of voices used throughout the fugue!

We have examined the summation of loudness of two or more superposed tones, but we have not yet discussed what happens to the threshold of hearing of a tone when it is sounded in presence of another. If the frequencies of both tones coincide, this threshold is given by the masking level

[5] A well-known situation, similar to this case, arises with the sensation of pain. If you are pinched in two places very near to each other, the pain is "twice" that of a single pinch (equivalent to case 1 above). But when the places are far apart, you have difficulty in sensing out what one may call "total pain" (case 3). Actually, you will tend to focus on only one of the two stimulations.

[6] Their addition, of course, also will contribute to a change in timbre (Chapter 4).

discussed earlier (p. 81). If their frequencies differ, we still can determine a masking level, defined as the minimum intensity level which the masked tone must exceed in order to be "singled out" and heard individually in presence of the masking tone. The intensity threshold of isolated pure tones (bottom curve in Fig. 3.12) changes appreciably, that is, increases, if other tones are simultaneously present. The most familiar experience of masking is that of not being able to follow a conversation in presence of a lot of background noise. The masking level ML of a pure tone of frequency f in presence of another pure tone of fixed characteristics (frequency 415 Hz and intensity level IL) is shown in Fig. 3.14 (Egan and Hake 1950). The level ML to which the masked tone has to be raised above the normal threshold of hearing (as given by the bottom curve of Fig. 3.12) is represented for different IL values of the masking tone. The regions close to f_0 must be extrapolated to the value deduced from the jnd of loudness (p. 81); beat phenomena play a role there, which has nothing to do with masking per se. At higher IL, additional complications arise due to the appearance of aural harmonics at frequencies $2f_0$, $3f_0$, etc. (Section 2.5). Notice the asymmetry of the curves at higher IL (caused by these aural harmonics): a tone of given frequency f_0 more efficiently masks higher frequencies than lower ones.

FIGURE 3.14. Masking level corresponding to a pure tone of 415 Hz, for various sound level values (Egan and Hake 1950) of the masking tone.

Reprinted by permission from the *Journal of the Acoustical Society of America.*

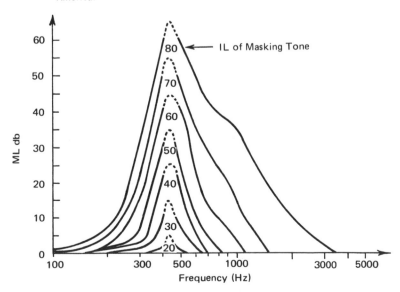

Masking plays an important role in polyphonic music, particularly in orchestration. On many occasions in musical scores, the participation of a given instrument such as an oboe or a bassoon may be totally irrelevant if it plays at the same time when the brasses are blasting a fortissimo. Likewise, the addition of flute stops or other soft string-like stops to a tutti of diapasons, mixtures, and reeds in the organ is completely irrelevant.

Finally, we must mention the effect of the *duration* of a tone on the loudness sensation. First of all, there is a *time threshold*—a minimum duration that a given pure sound must have in order to yield a tone sensation at all. This minimum duration is about 10–15 milliseconds, or at least 2–3 oscillation periods, if the frequency is below 50 Hz; tones lasting less than this are perceived as "clicks," not as "tones." Sounds lasting longer than 15 milliseconds (or 2–3 periods, whichever is longer) can be individualized as tones of given pitch and loudness.[7] Subjective loudness, however, does depend on tone duration (whereas pitch does not) (Plomp and Bouman 1959). The shorter the tone pulse, the lower its loudness, if the physical tone intensity (energy flow) is kept constant. When the tone duration is longer than about half a second, the loudness reaches a final value that is independent of duration (depending only on intensity).[8] Figure 3.15 shows a sketch of the relative loudness decrease, or loudness attenuation, as a function of tone duration, and for different frequencies. Notice that the "final" response value is reached sooner for the higher frequencies. This has important implications for music: if we want to play a "staccato" passage on the piano at a given loudness level, say "forte," we must hit the keys much harder than if we were to play the same passage "legato," at the same loudness level.[9] This effect comes out much more pronounced with nondecaying tones, such as organ sounds. The organist has a considerable control of the subjective loudness of one given note by giving it just the right duration (phrasing). Masking (Fig. 3.14), too, depends on tone duration, for short tones. The masking threshold increases as the tone duration decreases, for tones lasting less than about 0.5 sec. For these short tones it is found that, in first approximation, the loud-

---

[7] Quite generally, there is a physical principle which states that the frequency of a vibration cannot be defined more accurately than the *inverse of the total duration* of the vibration.

[8] For long exposure times an effect called *fatigue* sets in, decreasing again the subjective loudness sensation. The objective of trills is to prevent this from happening!

[9] For tones from musical instruments a complication arises: during tone buildup, which may last several tenths of seconds, a "natural" change in intensity and spectrum occurs at the source. Also, reverberation effects are important (Section 4.7).

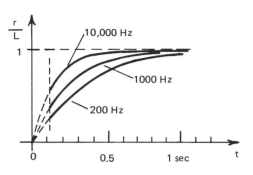

FIGURE 3.15

Relative loudness of pure tones of short duration. r/L: ratio of actual loudness (r) to loudness (L) of a steady tone of the same frequency and amplitude.

ness sensation is not related to power flow (intensity), but to total acoustical *energy* delivered to the ear (intensity $\times$ duration). Actually, there are indications that it is related to the total number of *neural impulses* transmitted in association with the short tone, rather than with energy (Zwislocki 1969) (see the next section).

3.5
The Loudness
Perception
Mechanism and
Related Processes

What physical or neural process is responsible for the difference between the limited subjective scale of loudness and the huge range of detectable intensities (Table 3.1; relation 3.17) of the original sound? In the case of primary pitch perception (Section 2.3) we already had encountered (without explicitly mentioning it) such a "compression": whereas the frequency scale ranges from about 20 Hz to about 16,000 Hz, this only comprises nine octaves of pitch. In that case the compression is mainly caused by the mechanical resonance properties of the cochlear partition: the curve in Fig. 2.8 indeed represents a roughly logarithmic relationship between the position x of the resonance region on the basilar membrane, and frequency.

In the case of the loudness detection mechanism, the "compression" is partly neural, partly mechanical. When a pure sound is present, the primary neurons connected to sensor cells located in the center of the region of maximum resonance amplitude increase their firing rate above the spontaneous level. This increase is a monotonic function of the stimulus amplitude, but not a linear one (See Chapter 4 of Flanagan 1972). Indeed, when the latter increases by a factor of, say, 100, (a 40 db increase in *SPL*, relation 3.16), the firing rate is found to increase only by a factor of 3 to 4. Another element contributing to loudness "compression" is related to the following: At higher *SPL*'s a primary neuron's firing rate *saturates* at a level that is only a few times that of spontaneous firing (the actual saturation level varying greatly from neuron to neuron). Any further increase in intensity would not alter greatly this firing rate; neurons simply cannot transmit pulses at a rate faster than the saturation value (determined by the refractory time after each pulse). How does one perceive a loudness increase even when the *SPL* surpasses the average neuron's saturation limit? What happens is that the more intense the sound wave is, the more extended the resonance region along the basilar membrane will be, hence the greater the *total number* of responding primary neurons whose thresholds are surpassed by the stimulus. In summary, an increase in intensity leads to an increase of the total number of transmitted impulses—either because the *firing rate of each neuron* has increased or because the *total number of activated neurons* has increased. This latter effect depends mainly on the shape of the membrane oscillation amplitude distribution—a purely mechanical property.

It follows from the previous discussion that the loudness sensation of a steady tone must be somehow related to the *total rate of neural impulses* triggered in the peripheral auditory system. As this neural information is transmitted to higher centers of the auditory tract and processed, the total firing rate gradually decreases (See Chapter 4 of Flanagan 1972). Typical acoustically evoked firing rates of cortical neurons are only 50–100 Hz. The relation between subjective loudness and total firing rate qualitatively explains the main properties of loudness summation (Section 3.4). For simultaneous tones of frequencies differing more than a critical band, the grand total of transmitted neural impulses is roughly equal to the sum of the pulse rates evoked by each component separately; hence, the total loudness will tend to be the sum of the *loudnesses* of each tone [relation (3.19)]. In contrast, for tones whose frequencies lie within a critical band, with resonance regions on the basilar membrane overlapping substantially, the total number of pulses will be controlled by the sum of the original stimulus *intensities* [relation (3.18)]. This picture may, however, be complicated by another type of loudness coding: nerve fibers innervating the outer rows of hair cells are excited by less intense sounds than those contacting the inner row (Section 2.8). In other words, there may be a component of loudness coding that appears in the form of a *spatial* distribution of neural activity—as in the case of primary pitch coding.

The dependence of loudness sensation (and masking thresholds) with tone duration (Fig. 3.15) points to a *time-dependent buildup of the acoustical signal processing operation*. Only after several tenths of seconds does the neural mechanism reach a "steady state" of tone processing. It is important to emphasize again that (quite fortunately for music) during this buildup the pitch sensation is stable and unique from the beginning (except for the very early stage).

There is, however, a slight dependence of pitch with loudness, for a tone of constant frequency. For tones above about 2,000 Hz, pitch increases when loudness increases and vice versa; below 1,000 Hz, the opposite happens (e.g., Walliser 1969). This effect is small; considerable changes in intensity are required to lead to noticeable changes in pitch.[10] This effect is probably caused by the asymmetry of the distribution of excitation along the basilar membrane and by the nonlinear neural response—a change in intensity causes a shift of the central point of excitation (even if the frequency remains constant), leading to a change in primary pitch sensation. Another result of this asymmetry may be the small, but rather relevant, effect in which the pitch of a pure tone of fixed frequency is found to shift slightly when another tone of different frequency is superposed (e.g., Walliser 1969, Terhardt and Fastl 1971). This effect may have some relevant consequences for musical intonation (Section 5.4).

Here we come again to the question, formulated on page 32, concerning the *primary* or spectral pitch detection mechanism—if a pure tone of given intensity and frequency sets the basilar membrane into resonant oscillation covering a finite spatial range Δx, and primary pitch is encoded in the form of spatial position x of the activated fibers, how come only *one* sensation of pitch is produced? This unique pitch sensation is believed to be partly the result of a *sharpening* mechanism in the peripheral neural network (Houtgast 1972). This mechanism operates in other sensorial systems too, being responsible for particularly important effects of contrast enhancement (e.g., Mach bands) in vision (Ratliff 1972). It probably works as follows. Impulses traveling up the afferent fibers that collect information from the hair cells lying in the stimulated region of the basilar membrane are branched off and conveyed to neighboring fibers through interneurons, which may act in either excitatory or inhibitory mode (Section 2.8). If the latter mode prevails, the transmission of signals from regions surrounding one of maximum stimulation is impaired, and the net result (after several stages of lateral interaction) would be a concentration or "funneling" of activity into a limited number of nerve fibers, surrounded by a ring of neural "quietude," or inhibition— ultimately leading to a unique sensation of primary pitch.[11] In general, this mechanism will enhance certain characteristic features of the spatial distribution of the original stimulus, such as maxima and abrupt changes.

---

[10] Pitch matching shows that an increase of SPL from 40 to 80 db causes the pitch of a 6,000 Hz tone to increase 7%, or that of a 150 Hz tone to decrease by 3% (Walliser 1969).

[11] It is possible that the efferent nerve bundle is the one that plays the inhibitory role in the sharpening mechanism (Section 2.9).

Finally, it should be mentioned that in the peripheral neural network of the visual tract, neurons have been discovered that respond only to *time changes* of the original light stimulus (Hubel 1971). As a similar behavior is expected for certain neurons in the auditory system, one might envisage the existence of a mechanism for direct coding of information on the *transient* characteristics of musical tones. This should be of importance to the perception of timbre (Chapter 4). A great deal of the corresponding neural processing may, indeed, already be done at the peripheral level of the cochlea (Lynn and Sayers 1970).

4 Generation of Musical Sounds, Complex Tones, and the Perception of Timbre

In the preceding two chapters the two principal tonal attributes pitch and loudness have been analyzed, mainly on the basis of pure, single-frequency tones. These are not, however, the tones that play an active role in music. Music is made up of *complex* tones, each one of which consists of a superposition of pure tones blended together in a certain relationship so as to appear to our ear as unanalyzed wholes. A third fundamental tonal attribute thus emerges: tone quality, or timbre, related to the kind of mixture of pure sounds, or harmonic components, in a complex tone (Section 1.2).

Most musical instruments generate sound waves by means of vibrating strings or air columns. In Chapter 1 we called these the primary vibrating elements. The energy needed to sustain their vibration is supplied by an excitation mechanism and the final acoustical energy output in many instruments is controlled by a resonator. The room or concert hall in which the musical instrument is being played may be considered as a natural "extension" of the instrument itself, playing a substantial role in shaping the actual sound that reaches the ears of the listener.

In this chapter we shall discuss how real musical tones are actually produced in musical instruments, how they are made up as superpositions of pure tones, how they interact with the environment in the rooms or halls, and how all this leads to the perception of timbre and instrument recognition.

4.1 Standing Waves in a String

We consider the case of a tense string, anchored at the fixed points P and Q (Fig. 4.1), of length L and mass per unit length d, stretched with a given force T that can be changed ad libitum, say, by changing the mass m of the body suspended from the string as shown in the figure. We now pluck or hit the string at a given point. Two transverse elastic

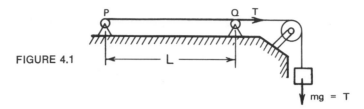

FIGURE 4.1

wave pulses will propagate to the left and to the right away from the region of initial perturbation in a manner discussed in Section 3.2, with a velocity given by relation 3.3. These wave pulses, when reaching the fixed anchor points P and Q will be reflected; a positive or "upward" pulse will come back as a negative or "downward" pulse, and vice-versa. After a certain time (extremely short, in view of the high speed of the waves in a tense string), there will be waves simultaneously traveling back and forth in both directions along the string. In other words, we will have elastic wave energy "trapped" in the string between P and Q, these two points always remaining at rest. If there were no losses, this situation would remain so forever and the string would continue to vibrate indefinitely. However, friction and leaks through P and Q will eventually dissipate the stored energy and the waves will decay.[1] For the time being we shall ignore this dissipation.

In view of the discussion in Section 3.3, we realize that the above picture of waves traveling back and forth along a string strongly resembles the situation arising in a standing wave. Indeed, it can be shown mathematically that *standing waves are the only possible stable form of vibration for a string with fixed ends, with the anchor points P and Q playing the role of nodes.*

This has a very important consequence. Among all imaginable forms of standing waves, only those are possible for which nodes occur at P and Q. In other words, only those sinusoidal standing waves are permitted that "fit" an *integer number of times* between P and Q (Fig. 4.2), i.e., for which the length of the string L is an integer multiple of the distance between nodes l_N given by relation (3.13). Taking into account this relation we obtain the condition $L = nl_N = n\lambda/2$, where n is any integer number 1, 2, 3, . . . This tells us that only the wavelengths

$$\lambda_n = \frac{2L}{n} \qquad n = 1, 2, 3, \ldots \tag{4.1}$$

---

[1] It is this energy "leak rate" through the "fixed" points (mainly the bridge) that is transformed into sound power in a string instrument's resonance body.

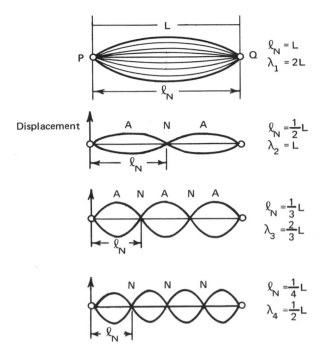

FIGURE 4.2. Standing wave modes in a vibrating string.

are permitted (Fig. 4.2). Using relation (3.8) we find that a string can only vibrate with the following frequencies:

$$f_n = \frac{1}{\lambda_n} \sqrt{\frac{T}{d}} = \frac{n}{2L} \sqrt{\frac{T}{d}} = n f_1 \tag{4.2}$$

The lowest possible frequency is obtained for $n = 1$:

$$f_1 = \frac{1}{2L} \sqrt{\frac{T}{d}} \tag{4.3}$$

This is called the *fundamental frequency* of the string. Note in (4.2) that all other possible frequencies are integer multiples of the fundamental frequency. They are called the upper *harmonics* of f_1 (Section 2.7). Notice in particular that the first harmonic ($n = 1$) is identical to the fundamental frequency; the second harmonic f_2 is the upper octave of f_1; the third harmonic is the twelfth (a fifth above the octave); the fourth harmonic the fifteenth (double octave); etc. (Fig. 4.3). The upper harmonics are also called *overtones*.[2]

Relation (4.3) tells us that the fundamental frequency of oscillation of a string is proportional to the square root of

---

[2] More precisely, *overtones* are the higher frequency components of a complex vibration, *regardless* of whether their frequencies are integer multiples of the fundamental frequency, or not.

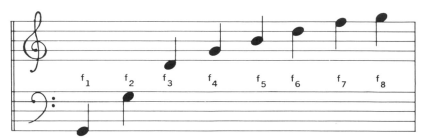

First Eight Harmonics of $f_1 = 98$ Hz (G_2)

FIGURE 4.3

the tension, that it is inversely proportional to its length and to the square root of its mass per unit length. This explains many characteristic features of piano strings: for the upper part of the keyboard, strings become shorter and shorter (higher fundamental frequency f_1); if we have to tune a given string a little sharper, we have to increase the tension (higher f_1) and conversely; in the low pitch region, in order to save space and maximize power output, instead of increasing the length of a string, its mass per unit length d is increased (lower f_1) by surrounding it with a coil of additional wire. In the violin, where we have only four strings of nearly equal length, each one must have different tension and/or mass in order to bear a different basic pitch. In order to vary the fundamental frequency f_1 of a given string, one changes its vibrating length L by pressing the string against the fingerboard, thus introducing a node at the point of contact.

The appearance in a natural way of frequencies that are integer multiples of a fundamental frequency receives the name of "quantization" and plays a fundamental role everywhere in physics. The different possible discrete forms of vibration of a physical system are called its *modes* of vibration. The fundamental, the octave, the twelfth, etc., are the first, second, third, etc., modes of vibration of a tense string. All frequencies of the possible vibration modes of a string are given by relation (4.2). In this relation only quantities depending on the string appear; *the vibration modes are thus a permanent characteristic of the particular physical system.* In which of the possible modes will a given string *actually* vibrate? This is determined by the way how the vibrations are initiated, i.e., by the primary excitation mechanism. Because of the capability of linear superposition of waves, *many different modes may occur simultaneously without bothering each other.* In this section we focus our attention on how a string *can* vibrate. In the next one we shall examine the question of how a string actually *will* vibrate.

Let us consider the case of a vibrating string in which only *one* mode is excited. This can be accomplished easily in the laboratory by letting an alternating electrical current of a given frequency flow through a tense metal string, spanned through the gap of a strong permanent magnet. Magnetic forces on the current in the string will drive a transverse vibration at the frequency of the current. Whenever this frequency is near one of the string's harmonics, a large standing wave is produced; one can visually observe the nodes and antinodes (as sketched in Fig. 4.2), and clearly hear the sound produced (provided the string is mounted on a resonating box). The use of a stroboscope (a light source that flashes at a given, controllable frequency) enables one to "freeze" the shape of the string, or to observe it in "slow motion."

A more accessible and widely known experiment can be done with a piano. Press slowly, and then hold down, the key of a bass note, say G_2 (Fig. 4.3), so that no sound is produced (the hammer does not strike) but the damper remains lifted off that string. Then hit hard and "staccato" the key of the first upper octave (G_3). After that sound has stopped, you very clearly hear the G_2 string vibrating one octave higher: it has been excited (by resonance) in its second harmonic mode, G_3! Now repeat the same experiment hitting the twelfth (D_4) while holding down G_2: you hear the G_2 string vibrating in D_4. Continue with G_4, B_4, and so on. As a test, hit A_3 or F_3 while holding down G_2—there will be *no* effect: the G_2 string remains silent. The reason is that A_3 or F_3 are not upper harmonics of G_2, and the G_2 string simply cannot sustain stable vibrations at those frequencies.

Relation (4.2) really is only an approximate one, particularly for the higher order modes. The reason is that the speed of a transverse wave in a string *does* depend slightly on the frequency (or wavelength) of the wave (this is called *dispersion*) and expressions (3.3) and (3.8) are not entirely correct. Indeed, the wave velocity V_T is slightly larger than the value given by relation (3.3). This deviation increases with increasing distortion of the string, i.e., it becomes more important for smaller wavelengths and larger amplitudes. The result is that the frequencies of higher vibration modes of a piano string are slightly sharper than the values given by (4.2).[3] In general, when the frequencies of the higher modes of vibration of a system are not integer multiples of the fundamental frequency, we call these modes *inharmonic*. Vibrating solid bodies other than strings—e.g., xylophone bars, bells, or chimes—have many inharmonic vibration

---

[3] This does not affect greatly the pitch of the ensuing complex tone (Appendix II). But it does affect tuning of the piano in the treble and bass register (when tuning is done by octaves).

modes, whose frequencies are not at all integer multiples of the fundamental frequency. In most of what follows, we shall assume, for simplicity, that the overtones of a vibrating string do coincide with the upper harmonics, and that relation (4.2) is exactly true. Thus we will often use indistinctly the terms "upper harmonics," "modes" or "overtones," although physically they are different concepts in the case of real strings.

4.2
Generation of
Complex Standing
Vibrations in
String Instruments

There are two fundamental ways of exciting the vibration of a tense string: (1) a one-time energy supply by the action of striking (piano) or plucking (harpsichord, guitar); and (2) a continuous energy supply by the action of bowing (violin family). In both cases, the resulting effect is a *superposition of many vibration modes activated simultaneously*. In other words, individual musical sounds generated naturally by strings contain many different frequencies at the same time —those of the vibrating system's harmonics. Figure 4.4 shows how this may arise in practice: adding the first and, say, the third harmonics, one obtains a resulting superposition that at a given instant of time, may look as depicted in this figure. Each mode behaves independently, and the instantaneous shape of the string is given by the superposition (sum) of the individual displacements, as dictated by each component separately. It is possible to use the previously mentioned experimental setup of a vibrating string in a magnetic field with an electric current flowing through it, this time using the combined output of *two* sinusoidal voltage generators whose frequencies are set equal to two of the string's harmonics, respectively. With the use of a stroboscope it is possible to clearly visualize the instantaneous shape of the string when it is vibrating in two modes at the same time. The relative proportion with which each overtone intervenes in the resulting vibration determines to a great extent (Sections 1.2 and 4.8) the particular character, quality, or timbre of the generated tone. The pitch of a string's complex tone is determined by the fundamental frequency (4.3), as has been anticipated in Section 2.7.

A simple experiment with a piano convincingly shows that a string can indeed vibrate in more than one mode at the same time. Press down slowly, and hold, a given key (say G_2) (Fig. 4.3) so to lift the damper off the string. Then hit hard and staccato *simultaneously* D_4, G_4, B_4. After their sound disappeared, it is possible to clearly hear the G_2 string vibrate in all three modes simultaneously—we have one string alone sound a full G major triad! What happens is that three modes (third, fourth, and fifth harmonics) have

FIGURE 4.4

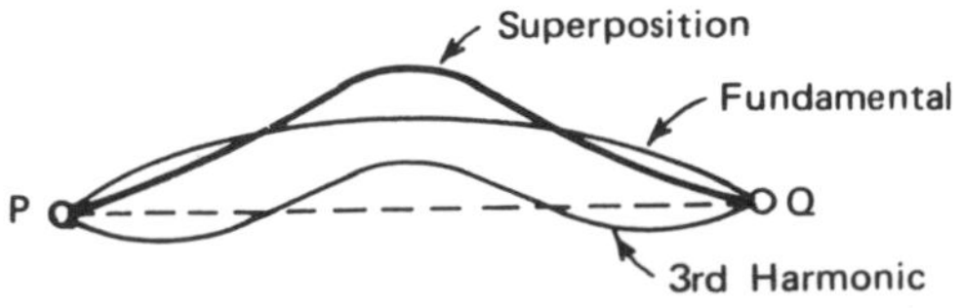

Superposition
Fundamental
P
Q
3rd Harmonic

been excited at roughly similar amplitudes (by resonance). A more drastic experiment is the following: hold the G_2 key down—and hit with your right underarm all black and white keys of two or more octaves before G_3—after the initial burst of noise has decayed, the G_2 string vibrates beautifully in the dominant seventh chord $G_3, D_4, G_4, B_4, D_5, F_5, G_5, \ldots$ (Fig. 4.3). Any of the other sounded notes could neither excite nor entertain a stable vibration on the G_2 string.

Whereas the above experiment shows that a given piano string *can* vibrate simultaneously in different modes at the same time, the following experiment shows that a piano string, sounded normally, actually *does* vibrate in many harmonic modes. Pick again a bass note, say, G_2. But this time press the G_3 key down slowly without sounding it, and keep it down. Then sound a loud, staccato G_2. The G_3 string starts vibrating in its own fundamental mode (i.e., G_3). The reason is that this mode has been excited (through resonance) by the *second* harmonic of the vibrating G_2 string. If instead of G_2 and had sounded A_2, the G_3 string would have remained silent. Then repeat the same experiment several times, successively pressing silently the keys of $D_4, G_4, B_4, D_5, \ldots$ etc. Each one of them will be excited by the corresponding upper harmonic mode of the G_2 string.[4]

Many vibration modes appear together when a string is set into vibration. What determines *which* ones and *how much* of each? This is initially controlled by the particular way the string is set into vibration, i.e., by the primary excitation mechanism. Depending on how and where we hit, pluck, or bow the string, different mixtures of overtones and, hence, different qualities of the ensuing sound will be obtained. We may explain this on the basis of the following examples. Assume that we give a string the initial form shown in Fig. 4.5a (this would be rather difficult to accom-

[4] The main objective of this *pedal* mechanism of the piano is based on this phenomenon: pressing the pedal lifts *all* dampers, and the strings are let free to vibrate by resonance. When one given note is sounded, all those strings will be induced to vibrate that belong to the series of harmonics of that note.

(a)

FIGURE 4.5

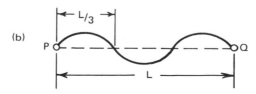

(b)

plish in practice, though). Since the shape more or less conforms to the fundamental mode (Fig. 4.2), the string will indeed start vibrating in that mode when it is released. If now the initial form is that shown in Fig. 4.5b, the string would vibrate in the third mode when released (Fig. 4.2). But what will happen if the initial shape has the far more realistic form of Fig. 4.6, which is obtained when we pluck the string at the midpoint A between P and Q? To find out let us superpose, i.e., add linearly, the cases of Figs. 4.5a and b. We obtain the shape shown in Fig. 4.7a which resembles rather well that of initial configuration of a plucked string (Fig. 4.6). We thus anticipate that the fundamental mode and at least the third harmonic should be simultaneously present in the vibration of a string plucked at the midpoint. We can improve greatly the approximation to the shape of Fig. 4.6 by adding more higher harmonics in appropriate proportions (Fig. 4.7b). One can iron out the remaining wiggles shown in this figure by just adding more and more higher harmonics in appropriate proportion until the wanted shape is almost exactly reproduced.

A remarkable fact is that there is no guesswork involved in all this: it can be accomplished in rigorously mathematical form! In fact, it can be shown that *any arbitrary initial shape of a string can be reproduced to an arbitrary degree of accuracy by a certain superposition of geometrical shapes corresponding to the string's harmonic vibration modes* (standing waves). It is this "mathematical" superposition of shapes, in particular, the proportion of their amplitudes and phases, which definites the physical superposition of harmonics with which the string will actually vibrate when it is released from its initial configuration. In other words, each one of the component standing waves, which when added together make up the initial form of the string (e.g., Fig. 4.7b), proceeds vibrating in its own way with its own characteristic frequency and amplitude once the string is released. As time goes on, the instantaneous form of the string changes periodically in a complicated way; but every time a fundamental period $\tau = 1/f_1$ has elapsed, all component modes will find themselves in the same relationship as in the beginning, and the string will have the same shape as it had initially. It is very important to remark here that the initial configuration of the string determines not only the amplitudes of the harmonic vibration modes but also their phases (relative

FIGURE 4.6

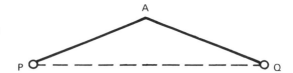

(a)

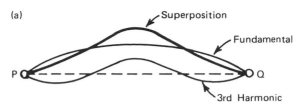

FIGURE 4.7

(b)

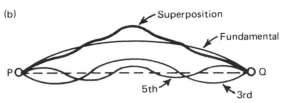

timings). The point at which the string is plucked will determine the particular proportion of upper harmonics, i.e., the initial timbre of the sound emitted (Section 1.2). If we pluck at the center, we will have the situation shown in Fig. 4.7b, and only *odd* harmonics will appear. On the other hand, the closer to the fixed extremes we pluck, the richer the proportion of upper harmonics will be. In general, all those harmonics that have a node at the plucking point will be suppressed (e.g., all even harmonics in the example of Fig. 4.6), whereas those having an antinode there will be enhanced. This effect is most efficiently exploited by the harp player to control the timbre of his instrument's sound.

In a string that is set in vibration by plucking, we have a situation in which the primary excitation mechanism initially gives a certain *potential energy* to the system, by deforming the string. After release, this initial energy is periodically converted back and forth into kinetic energy of vibration (Section 3.1). On the other hand, when a string is set into vibration by striking a certain amount of *kinetic energy* is initially provided by the striking mechanism (e.g., the hammer in the piano), setting the points of the initially undeformed string in motion. This initial energy is then periodically converted into potential energy of deformation. It can be shown mathematically that *from the knowledge of the initial velocities of the points of the struck string, the ensuing superposition of harmonics can be deduced.* Thus, a string hit at the midpoint will oscillate principally with the fundamental frequency, plus a mixture of decreasing intensities of the odd harmonics. The closer to the end points P or Q a string is struck, the richer in upper harmonics the tone will be. As it happened with the plucked string, harmonics whose nodes are at or near the striking point will be excluded, and those having an antinode there will be enhanced.

When a string is set into vibration by plucking or striking, we observe the vibration to decay away quite rapidly. This is caused by the action of dissipative forces: elastic friction inside the string and, most importantly, forces that set into small vibratory motion whatever is holding the string in place at its end points. Only part of this energy loss is actually converted into sound wave energy. A freely vibrating string mounted on a rigid, heavy frame produces only a faint sound: most of the vibration energy disappears forever in form of frictional energy (heat). The conversion into sound wave energy can be greatly increased by mounting the string on a board of special elastic properties, called a *resonator* (sound board in the piano, the body of a violin). In that case the end points of the string are allowed to vibrate a tiny bit (so little that, as compared to the rest of the string vibrations, these end points technically still function as nodes), and the energy of the string can be gradually converted into vibration energy of the board. Owing to the usually quite large surface of this board, this energy is then converted with greater efficiency into sound wave energy. The resulting sound is much louder than in the case of a rigidly mounted string—but *it decays much faster*, because of the considerably increased *rate* at which the available amount of string energy is spent (power, Section 3.1).

Let us examine the process of vibration decay in more detail. For simplicity we consider a string that has been set into free vibration in its fundamental mode only. We focus our attention on the gradually decreasing amplitude of the oscillation of the string, say, at an antinodal point. Measurements show that, for a given string, damped oscillations with larger amplitude decay at a faster rate than those with smaller amplitude. The resulting motion is shown in Fig. 4.8. Notice the slope of the "envelope" curve, which decreases as the amplitude decreases. This is called an *exponential* decay of amplitude. Most important (and quite fortunately for music!), the frequency of a damped oscillation remains constant.

This is roughly the way a string behaves when it vibrates freely in one given mode after it is struck or plucked. If it is mounted on a rigid board the energy loss will be relatively small and so will be the amplitude damping (Fig. 4.9a). If, instead, it is mounted on a sound board, it will give away energy at a larger rate by setting the board and the surrounding air into oscillation. The oscillations will thus decay faster (Fig. 4.9b).

A characteristic quantity is the so-called *decay half-time*. This is the time interval after which the amplitude of the oscillations has been reduced to one-half the initial value (Fig. 4.8). The remarkable fact of an exponential decay is that this half-time is always the same throughout the decay: it takes the same time to halve the amplitude, no matter what the actual value of the latter is. The decay half-time is

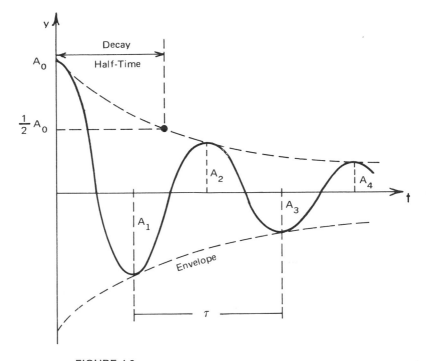

FIGURE 4.8

thus a characteristic constant of a damped oscillation. The typical decay half-time of a piano string is about 0.4 seconds.

When a string vibrates in several modes at the same time, the situation is more complex. However, we still find that *each mode* decays exponentially, only that the decay half-time will be different for different modes. The resulting complex sound thus not only decreases in loudness, but also its timbre will gradually change. In the piano strings the

FIGURE 4.9

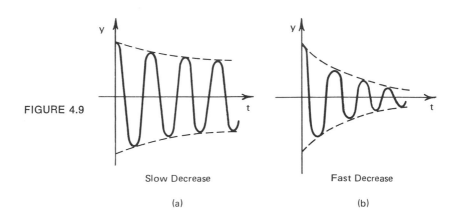

Slow Decrease   Fast Decrease

(a)   (b)

upper frequency modes decay somewhat faster than the lower harmonics; in a vibrating bell the lower harmonics continue to sound long after the upper ones have decayed away. Otherwise, the overall behavior of a freely vibrating string is exclusively determined by the way in which the vibration has been initially excited (hit or plucked string). In the piano, for instance, this is controlled exclusively *by the speed with which the hammer hits*—nothing else. In the case of a *single* tone the so-called piano "touch" is nothing but plain loudness with a timbre that is irrevocably coupled to that loudness and to the subsequent decay. Tone quality cannot be controlled independently. The "touch" that undoubtedly exists when a musical piece (i.e., a sequence and or superposition of tones) is played is related to other psychoacoustical effects (mainly tone *duration control*, small variations of loudness from tone to tone, and *nonsimultaneity* of the different notes of a chord).

What can we do in order to avoid the damping of a string vibration? Obviously, we must compensate the energy loss by somehow supplying extra energy to our vibrating system at a rate equal to the dissipated power. If the supplied power *exceeds* the energy loss rate by a certain amount, the amplitude will gradually build up. But this buildup would not go on indefinitely: while the power supply remains constant, the power dissipation will increase as the amplitude increases, and a regime will eventually be attained in which the dissipated power has become equal to the supplied power (Fig. 4.10). This happens during the tone buildup of any instrument of continuous sounding capability (bowed violin string, flute, organ pipe, etc.). In such a case each harmonic is found to build up independently, as if there were an individual power supply mechanism for each mode. The larger this power supply, the larger, of course, will be the final intensity level.

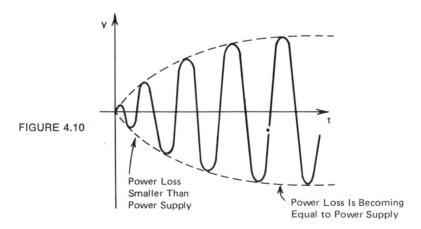

FIGURE 4.10

The bowing mechanism is a good example of how string oscillations can be sustained in a constant regime. The physical problem is mathematically complicated and can be treated only after making several simplifying assumptions (Friedlander 1953; Keller 1953). Here we can only give a qualitative description of the theory. The interaction between the bow and the string is produced by forces of friction. Quite generally, we distinguish two types of frictional interaction. The first is so-called *static* friction that arises when there is *no displacement* between the interacting bodies. This happens when the string "sticks" to the bow, thus moving with the same speed as the latter (or, in more familiar terms, when you try to push a heavy table while it "sticks" to the floor). The second type is *dynamic* friction, arising when the two interacting bodies (their contact surfaces) are *sliding* against each other. This happens when the string "snaps back" and moves in the opposite direction as that of the bow (and happens while you continue to push a ·table *after* it had started moving). Dynamic friction is weaker than static friction; both mechanisms are controlled by the force, perpendicular to the surface of contact, with which one body presses against the other (and vice versa). In the case of a bowed string, this perpendicular force is called the *bowing pressure*—a dreadful name in the ears of a physicist, because it is *not* a pressure, just a force.

In Appendix I we discuss in more detail an idealized situation. The main physical conclusions are as follows: (1) *the amplitude of the vibration of a bowed string (loudness of the tone) is controlled solely by the bow velocity, but in order to maintain constant the nature or type of the string motion (timbre of the tone), one must keep the bowing pressure proportional to the bowing speed.* This is well known by string instrument players, who increase both bow speed b and bowing pressure P at the same time to produce an increase in loudness *without* a change of timbre, or who increase b and *decrease* P, to produce an increase in loudness *with* change in timbre. (2) *A bowed string always has an instantaneous shape that is made up of sections of straight lines;* this result was verified experimentally long ago. A study of the energy balance in the bowing mechanism reveals that most of the energy given to the string by the bow during the "sticking" portions of the motion is spent in the form of frictional heat (work of the dynamic friction force) during the slipping phases. Only a small fraction is actually converted into sound energy![5]

As in the case of a plucked or struck string, the particular mixture of harmonic modes of vibration will depend on the

---

[5] For a detailed discussion of the bowing of strings, see Schelleng (1973).

position of the bowing point. Bowing close to the bridge (*sul ponticello*) will enhance the upper harmonics and make the tone "brighter"; bowing closer to the string's midpoint (*sul tasto*), reduces the intensity of upper harmonics considerably, and the sound is "softer."

In the previous discussion we have tacitly assumed that the bow is being displaced exactly perpendicular to the string. If it has a small component of parallel motion, *longitudinal* vibration modes of the string can be excited. Their frequency is much higher than the fundamental frequency of the transverse modes, longitudinal oscillations are responsible for the squeaky sounds heard in beginners' play.

4.3
Sound Vibration
Spectra and
Resonance

When a string vibrates in a series of different modes at the same time, the generated sound waves are complex, too. Each harmonic component of the original string vibration contributes its own share to the resulting sound wave, of frequency equal to that of the corresponding mode and of intensity and phase that are related to the intensity and phase of the latter through the intervening transformation processes. The result is a superposition of sound waves blended together into one complex wave, of fundamental frequency f_1 [equal to the fundamental frequency of the vibrating element (4.3)] and with a series of upper harmonics of frequencies $2f_1$, $3f_1$, $4f_1$, ... etc. The resulting vibration is periodic, repeating with a period equal to $\tau_1 = 1/f_1$. In other words, the fundamental frequency f_1 also represents the repetition rate of the resulting complex vibration (Section 2.7). The shape of the resulting curve depends on *what* harmonics are present, on *how much* there is of each (i.e., on their relative amplitudes), and on their *relative timing* (i.e., their relative phases).

And here we come to a mathematical theorem that has had an absolutely smashing impact on practically every branch of physics—in particular, physics of music. In short, the theorem states that *any periodic vibration*, however complicated, *can be represented as the superposition of pure harmonic vibrations*, whose fundamental frequency is given by the repetition rate of the periodic vibration. But that is not all: this theorem also provides all mathematical "recipes" for the numerical determination of the amplitudes and phases of the upper harmonic components! It is called Fourier's theorem, named after a famous nineteenth-century French mathematician. The determination of the harmonic components of a given complex periodic motion is called *Fourier analysis*; the determination of the resultant complex periodic motion from a given set of harmonic components is called *Fourier synthesis*. Similarly, given a complex tone, the process of finding the harmonic components is called *sound analysis*. Conversely, given a group of harmonic com-

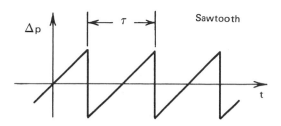

FIGURE 4.11

ponents, the operation of blending them together to form a complex tone is called *sound synthesis*.

Let us discuss one example of Fourier analysis. Of course, we cannot give here the whole mathematical operation that is needed to obtain the numerical results; we will have to accept these and, rather, concentrate on their physical interpretation. We pick the periodic motion shown in Fig. 4.11, corresponding to a "saw-tooth" wave. τ is the period, $f_1 = 1/\tau$ the repetition rate or fundamental frequency. This type of vibration can be generated electronically. To some extent it represents in an idealized form the motion of a bowed string. Figure 4.12 shows how this periodic motion can be made up

FIGURE 4.12. Fourier analysis (up to the sixth harmonic) of a sawtooth wave.

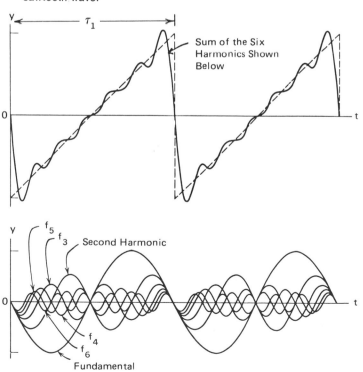

of pure, harmonic vibrations. Of course, many wiggles still remain in the curve corresponding to the sum. But this is because, for better clarity, we have stopped adding after only the sixth harmonic. Adding more and more upper harmonics (of amplitude and phase obtained by the method of Fourier analysis) will iron out these wiggles, and the saw-tooth shape will be approached more and more closely. Observe attentively the relative amplitudes and the relative timing of the harmonic components, and how positive and negative portions add up to give the resultant curve. With a trained eye and intuition it is possible to predict qualita-tively the principal harmonic components of a periodic motion of almost any, arbitrary, shape.

We now have to find a way to characterize physically a given complex sound. In principle, we must specify *three distinct series of quantities:* the frequencies of the harmonic components, the pressure variation amplitudes or intensities of the components, and their phases or relative timings (e.g., Fig. 2.5). In practice, however, it is customary to specify only the fundamental frequency f_1, and the intensities of the harmonic components, because, first, it is understood that all upper frequencies are just integer multiples of the funda-mental frequency f_1, and second, the phases of the com-ponents play only a secondary role in timbre perception, particularly for the first (and most important) half-dozen or so harmonics (Section 4.8).

The sequence of intensity values $I_1, I_2, I_3, \ldots$, of the harmonic components of a complex tone represents what is called the *power spectrum* of the tone. Two complex tones of the same pitch and loudness but different spectrum sound differently, i.e., have a different tone quality. The difference in spectrum gives us an important cue to distinguish between tones from different instruments—but other cues, partic-ularly tone attack and decay, are also needed for instrument identification (Section 4.9). The fact that a multiplicity of physical parameters $(I_1, I_2, I_3, \ldots)$ is related to timbre indicates that the latter is a *multidimensional* psychophysical magnitude.

Tone spectra can be represented graphically by plotting for each harmonic frequency (horizontal axis) the intensity with which that harmonic component intervenes (vertical axis) (e.g., Fig. 4.16). Quite often, values of IL (3.15) or SPL (3.16) are used instead of intensities to represent a spectrum. Also, intensity or IL values *relative* to that of the fundamen-tal, or relative to the total intensity $I = I_1 + I_2 + I_3 + \ldots$, are used. There are many books in which sound spectra of actual musical instruments are reproduced (e.g., Culver 1956). A word of caution is necessary, though. From a psychophysical point of view, a conventional harmonic

(Fourier) representation of a tone spectrum makes no real sense beyond the sixth or seventh harmonic, because in that range neighboring components start falling within a critical band. Since this is the elementary acoustical information-collection and -integration unit of the ear (Section 2.4), the auditory system could not resolve the individual intensities of these upper harmonics [case (1) on p. 85]. A psycho-physically more meaningful representation of tone spectra is obtained by listing the integrated intensity values *per critical band* (frequency intervals of roughly $\frac{1}{3}$-octave extension).

Only steady tones can be "resolved" into a superposition of harmonics of discrete frequencies that are integer multiples of the fundamental. When a vibration pattern is *changing* in time, this is not possible anymore. However, a sort of "expanded version" of Fourier analysis is applicable. It can be shown mathematically that a time-dependent tone leads to a *continuous spectrum* in which all frequencies are represented, with a given intensity for each infinitesimal frequency interval. If the tone is *slowly* time-dependent, discrete frequencies (those of the harmonics) still will be represented with highest intensity (spectral "peaks"), but if the time change is appreciable from cycle to cycle, the discrete character will disappear and the spectrum will tend to become a continuous curve covering the full frequency range (even if the tone was originally pure). This fact leads to another important observation concerning "high fidelity" reproducing equipment (see also p. 37). We noted in Section 1.2, and will come back to this in Section 4.8, that *transients*, i.e., rapid time variations of the vibration pattern of a tone play a determining role in the perception of quality or timbre. Thus, in order to reproduce the transients of a given tone correctly, the recording and reproducing systems must leave the tone spectrum undistorted *over the full frequency range*. Our ear does not need the frequency components lying much over about 5,000 Hz of a steady tone, but the reproducing system needs them to give us a correct version of the rapidly changing portions of a tone!

The sound spectrum of a string instrument is not at all equal to that of the vibrations of the strings. The reason lies in the frequency-dependent efficiency of the *resonator* (sound board of a piano, body of the violin), whose main function is to extract energy from the vibrating string at an enhanced rate and convert it more efficiently into sound wave power. As mentioned earlier, the string vibrations are converted into vibrations of the resonator in a process in which the "fixed" end points of the string (particularly the one situated on the bridge) are allowed to vibrate a tiny bit. This remnant vibration is so small that it does not invalidate the fact that, from the string's point of view, these points still are vibration nodes. In spite of being so small, these vibrations *do* involve an appreciable energy transfer.[6] The explanation is found in the very definition of work (Section 3.1): although the displacement of the string's end points is extremely small, the *forces* applied on them are large (of the order of the tension of the string), so that the *product* force times displacement (work) may be quite appreciable. Owing to the large surface of a typical resonator, conversion of its vibration energy into sound wave energy is very efficient—thousands or even millions of times more efficient than the direct conversion of a vibrating string energy into sound.

Like a string, the complex elastic structure of a piano soundboard or body of a violin has preferred modes of oscillation. In this case, however, there is no such simple integer multiple relationships between the associated frequencies as given in relation (4.2). Moreover, there are so many modes with nearly overlapping frequencies that one obtains a whole continuum of preferred vibration frequencies, rather than discrete values.[7] Let us briefly discuss how these vibration modes arise. To that effect, instead of considering the violin body as a whole, we examine the vibration of just one of the plates of the body. To find out the possible vibration modes it is necessary to excite the plate with a sinusoidal, single-frequency mechanical vibrator, at a given point of the plate (e.g., with which the bridge is normally in contact). Elastic waves propagate in two dimensions away from the excitation point and are reflected at the edges of the plate. The only stable modes of vibration are standing waves compatible with the particular boundary conditions of the plate. This process is highly complicated, impossible to treat mathematically. In the laboratory, however, it is possible to make the vibrations of the plate visible through a modern laser technique called holography (Reinicke and Cremer 1970). The simplest mode of oscillation (called "ring mode") is one in which the central region of the plate moves sinusoidally up and down, with the boundary acting as a nodal line. The ring modes of the violin plates determine the "tap tone," the sound that is evoked by tapping the body. Figures 4.13 and 4.14 (Jansson et al. 1970) depict holograms of four successive vibration modes of the top plate (with *f*-holes and sound post, but without fingerboard) and back plate of a violin, respectively. Each one of the dark curves represents a contour of equal deformation amplitude. The difference between neighboring fringes is about 2×10^{-5} cm.

---

[6] The function of the *mute*, when it is applied to the bridge of a string instrument, is to decrease this energy transfer for the higher frequency components, thus altering the quality of the resulting tone.

[7] Neither does a *real* string, of *finite* thickness, have "sharp," discrete modes of oscillation.

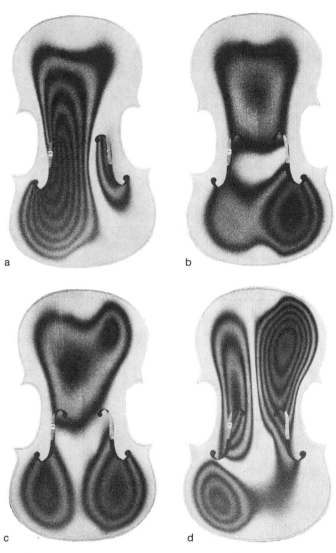

FIGURE 4.13. Holograms depicting the first four vibration modes of the top plate of a violin (with *f*-holes and mounted sound post, without fingerboard). Each one of the dark curves represents a contour of equal deformation amplitude. (a) 540 Hz; (b) 775 Hz; (c) 800 Hz; (d) 980 Hz.

Reprinted by permission from Jansson et al. 1970.

In the *assembled* instrument, the vibration modes of the top plate (Fig. 4.13) remain nearly unchanged, but new vibration modes appear (in the low frequency range).

The vibration response of a resonator to a given signal of *fixed* amplitude (either from a mechanical vibrator, or from a vibrating string mounted on that resonator) depends

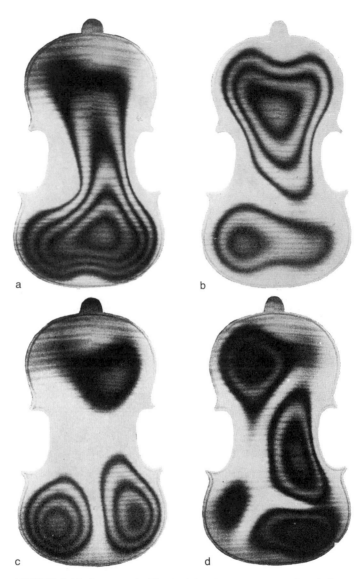

FIGURE 4.14. Same as in Fig. 4.13 for the back plate of a violin.
(a) 740 Hz; (b) 820 Hz; (c) 960 Hz; (d) 1,110 Hz.

Reprinted by permission from Jansson et al. 1970.

strongly on the frequency of the primary oscillations. For
that reason, a soundboard reacts differently to vibrations of
different frequency. Some frequencies will be preferentially
enhanced, whereas others may not be amplified at all. A
frequency for which the energy conversion is especially
efficient is called a *resonance frequency* of the resonator.
A resonator may have many different resonance frequencies;
they may be well defined ("sharp" resonance) or spread

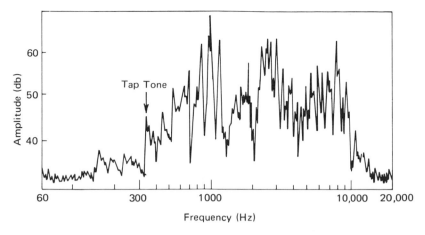

FIGURE 4.15. Resonance curve of a violin plate (Hutchins and Fielding 1968).

Permission from *Physics Today* is acknowledged.

over a broad range of frequencies. The graph obtained by plotting the output signal (for instance, as measured by the intensity of the emerging sound wave) as a function of the frequency of a sinusoidal input vibration of constant amplitude, is called a *resonance curve*, or response curve. Usually, the intensity of the output signal I is represented in relationship to some given reference signal I_{ref}, and expressed in *decibels* (db) (Section 3.4):

$$R = 10 \log \frac{I}{I_{ref}} \quad \text{(in decibels)} \tag{4.4}$$

where R is the value of the response function. The dependence of R with frequency gives the above-mentioned resonance curve. Figure 4.15 is an example, corresponding to the plate of a violin (Hutchins and Fielding 1968). The first sharp rise marked in Fig. 4.15 corresponds to the tap tone.[8] The response curve of an *assembled* violin shows a first resonance peak in the 280–300 Hz range (near D string pitch!) corresponding to the first vibration mode of the *air* enclosed by the body. The next resonance, usually about a fifth above the air resonance, is called the *main wood resonance*. Beyond 1,000 Hz, the multiple resonance peaks of an assembled instrument are considerably less pronounced than those of a free plate, shown in Fig. 4.15. The response curve of a piano soundboard is even more complicated; this complication, however, ensures a relatively even amplification over a wide range of frequencies.

[8] The position (in frequency) and the shape of this particular resonance peak is of capital importance for the quality of a string instrument (Hutchins and Fielding 1968). See also p. 142.

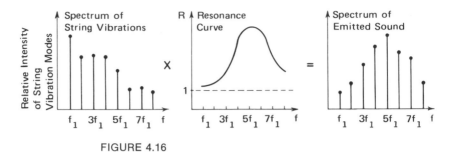

FIGURE 4.16

Figure 4.15 represents the response curve of a resonator to a single, *harmonic* vibration of given frequency f. What happens if it is excited by a string vibrating with a complex spectrum of harmonics, or frequencies f_1, $f_2 = 2f_1$, $f_3 = 3f_1$, ... etc., and intensities I_1, I_2, I_3, ... ? Each harmonic component will be converted independently, as prescribed by the value of the response function R corresponding to its own frequency. The timbre of the resulting sound is thus governed by both, the original string vibration spectrum *and* the response curve of the resonator.

As an example, consider the hypothetical spectrum of a vibrating string shown in Fig. 4.16. This string is mounted on a hypothetical soundboard the response curve of which is also shown. The output sound spectrum is given in the right graph (relative values). The fundamental is considerably reduced; instead, the fifth harmonic appears enhanced above all others. According to Fig. 4.16, in this example more power would be extracted from the fifth harmonic than from any other. If the string had originally been plucked or struck, this harmonic would decay faster than the others because its energy reservoir would be depleted faster. This leads to a time-dependent change in spectrum or tone quality as the sound dies away. If, on the other hand, the string were to be bowed, the energy loss in each mode would automatically be compensated for by the bowing mechanism, the resulting tone quality remaining constant in time.

Finally, we come to a point most important for music. The response curve of a resonator is an immutable characteristic of a musical instrument. If, for instance, it has a resonance region around, say, 1,000 Hz, it will enhance all upper harmonics whose frequencies fall near 1,000 Hz, no matter *what* note is being sounded (provided, of course, that its fundamental frequency lies below 1,000 Hz) and no matter what the original string vibration spectrum was. A broad resonance region that enhances the upper harmonics lying in a fixed frequency range is called a *formant*. A musical instrument (its resonator) may have several formants. It is believed that formants, i.e., the enhancements of harmonics in certain fixed, characteristic, frequency intervals are used by the auditory system as a most important "signature" of a complex tone in the process of *identification* of a musical instrument (Section 4.9). One of the reasons in favor of this hypothesis is the fact that formants are the only invariable characteristic common to most, if not all, tones of a given instrument, whereas the spectrum of individual tones may vary considerably from one note to another.

4.4
Standing
Longitudinal Waves
in an Idealized
Air Column

Let us consider a long, very thin cylinder, open at both ends (Fig. 4.17). The air inside can be considered as a unidimensional elastic medium (Section 3.2) through which longitudinal waves can propagate. At any point *inside* the cylinder the pressure is allowed to momentarily build up, decrease, or oscillate considerably with respect to the normal atmospheric pressure outside—the rigid walls and the inertia of the remaining air column hold the necessary balance to the forces (3.1) that arise because of the pressure difference. But at the open end points P and Q, no large pressure variations are allowed even during the shortest interval of time, because nothing is there to balance the arising pressure differences. These points thus must play the role of *pressure nodes,* and any sound wave caused by a perturba-

FIGURE 4.17. Standing wave modes (pressure variations) in an idealized cylindrical pipe, open at both ends.

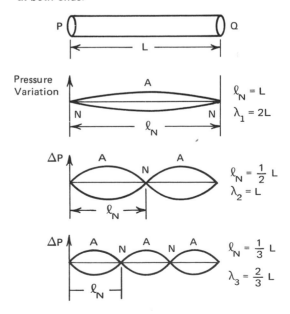

tion inside the pipe and propagating along it will be reflected at either open end. We hence have a situation formally
analogous to the vibrating string, discussed in Section 4.1:
sound waves generated in the pipe remain trapped inside
and *the only possible stable vibration modes are standing
longitudinal waves with pressure nodes at the open ends
P and Q* (Fig. 4.17). Notice that in view of our discussion
in Section 3.3 on page 77, the open end points are *antinodes of displacement,* i.e., points with maximum vibration
amplitude.

The open air column does not necessarily have to be
physically defined in the manner shown in Fig. 4.17. For
instance, there is an open air column comprised between
points P and Q of the pipe shown in Fig. 4.18. Indeed, since
there are holes at P and Q, the air pressure at these points
must remain constant and equal to the external pressure.
P and Q thus play the role of open ends of the enclosed
air column. Figure 4.18 corresponds to the case of an idealized flute, where P is the mouthhole and Q the first open
fingerhole.

In a real open pipe of finite diameter, *the pressure nodes
do not occur exactly at the open end,* but at a short distance further out ("end correction," p. 126). The relations
given below are thus only first approximations.

From Fig. 4.17 and relation (3.6) we obtain the frequencies of the vibration modes of an open cylindrical pipe:

$$f_n = \frac{n}{2L} \, 20.1 \, \sqrt{t_A} = n \, f_1 \qquad n = 1, 2, 3, \ldots . \tag{4.5}$$

f_1 is the fundamental frequency

$$f_1 = \frac{10.05}{L} \sqrt{t_A} \tag{4.6}$$

Remember that t_A is the *absolute* temperature of the air in
the pipe, given by (3.5). L in (4.5) and (4.6) must be expressed in *meters.* Taking into account that the wavelength
λ_1 of the fundamental tone is related to the length L of the
tube by $\lambda_1 = 2L$ (Fig. 4.17) and inspecting Fig. 3.7, one may
obtain an idea of typical lengths of open flue organ pipes,
flutes, and recorders as a function of frequency. An increase in frequency (pitch) requires a decrease in length. Relation (4.6) also shows the effect of air temperature on the
fundamental pitch of a vibrating cylindrical air column. An

FIGURE 4.18

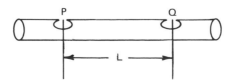

increase in temperature causes an increase in frequency (sharper tone). Thus flutes and flue organ pipes must be tuned at the temperature at which they are expected to be played. Fortunately, the fundamental frequency (4.6) is controlled by the absolute temperature t_A, appearing under a square root. Both facts make the influence of temperature variations on pitch a rather weak one, but enough to be concerned with—as flautists and organists well know.

We now turn to the case of a narrow cylinder closed at one end (Fig. 4.19). We realize that, whereas at the open end P the pressure must remain constant and equal to that of the outside air (pressure node), at the stopped end Q the inside pressure can build up or decrease without restriction. Indeed, a *pressure antinode* is formed at A. This is more easily understood by considering the actual vibratory motion of the points of the medium. Quite obviously there must be a *vibration node* for all air molecules near Q: they are prevented from longitudinal back-and-forth oscillation by the cover of the pipe. According to the discussion in Section 3.3, such a vibration node corresponds to a pressure antinode.

Figure 4.19 shows how the standing wave modes "fit" into a stopped pipe in such a way as to always have a pres-

FIGURE 4.19. Standing wave mode in an idealized cylindrical pipe, closed at one end.

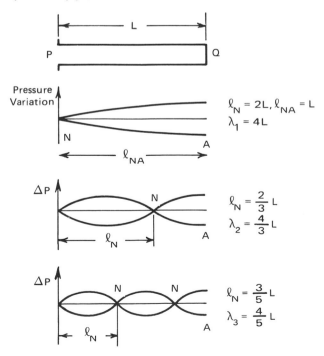

sure node at the open end and a pressure antinode at the stopped end. For the fundamental frequency we find the relation

$$f_1 = \frac{1}{4L} 20.1 \sqrt{t_A} = \frac{5.03}{L} \sqrt{t_A} \tag{4.7}$$

[L in meters, t_A absolute temperature (3.5)]. This is exactly *half* of the fundamental frequency (4.6) of an open pipe of the same length. In other words, an idealized stopped cylindrical pipe sounds an octave below the pitch of a similar pipe open at both ends.

With respect to the higher modes of a stopped cylindrical pipe, an inspection of Fig. 4.19 (and conversion of wavelength to frequency) reveals that only *odd* multiples of the fundamental frequency f_1 (4.7) are allowed:

$$f_1, f_3 = 3f_1, f_5 = 5f_1, \ldots \tag{4.8}$$

The frequencies $2f_1, 4f_1, 6f_1, \ldots$ are forbidden—their modes cannot be stably sustained in an ideally thin stopped cylindrical pipe. In other words, *the overtones of a stopped pipe are odd harmonics of its fundamental.*

The clarinet is perhaps the most familiar example of an instrument behaving very nearly like a stopped cylindrical pipe. The mouthpiece with the reed behaves as the closed end, the bell or the first open fingerhole defining the open end. The fundamental pitch of a note played on the clarinet indeed lies one octave below the note corresponding to the same air column length, played on a flute.

Organs include several stopped pipe ranks. One of the reasons is money and space savings: open bass pipes are very long (according to relation (4.6) an open C_1 pipe is 5.3 m high). The same pipe, if stopped, needs only to be 2.7 m long. Of course, there is more than money at stake: a stopped pipe yields a different quality of sound than an open one of the same fundamental frequency.

Our last case under discussion here is a (very narrow) conical pipe, stopped (closed) at the tip P (Fig. 4.20). The determination of the modes of vibration requires a rather complex mathematical analysis. The results can be summarized rather simply: an idealized narrow conical pipe stopped at the tip has the same vibration modes as an open pipe of the same length. In other words, relations (4.5) and (4.6) apply. A *truncated* (narrow) cone (Fig. 4.21), closed at

FIGURE 4.20

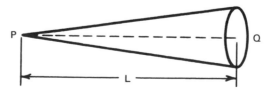

FIGURE 4.21

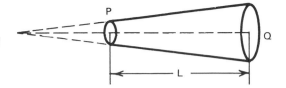

the end P, has a series of vibration modes that do not bear the integer number relationship: in the lower frequency range (near the fundamental) they correspond nearly to the modes of an open pipe of the same length L, but for higher frequencies they approach those of a closed cylindrical pipe of length L. In other words, the vibration modes are *inharmonic*.

4.5
Generation of
Complex Standing
Vibrations in
Wind Instruments

The results of the preceding section are idealizations that would become true only for hypothetical air columns (cylinders, cones) of diameters that are very, very small as compared to their length L. This, however, is not the case with real musical instruments and organ pipes. Besides, the air cavities in these instruments are cylindrical or conical only along a certain portion of their length, with more complex shapes near the mouthpiece and the open end (open fingerholes, bells, etc).

In order to analyze the physical behavior of real wind instruments we must inspect in more detail all phenomena involved. Let us first turn to the *excitation mechanism*. There is no equivalent to the "plucking" or "hitting" of a string in the case of an air column. The reason is that the vibrations of a freely oscillating air column decay almost instantaneously. One may verify this easily by tapping with the hand on one end of an open pipe (the longer the better), or by knocking sharply on its wall, while holding the ear near the other end. A sound burst of pitch equal to the fundamental frequency of the pipe can indeed be heard, but the decay takes place within a fraction of a second. It is thus necessary to have a primary excitation mechanism equivalent to the bowing of a string, that continuously supplies energy to the vibrating air column at a given rate.

There are two distinct types of such mechanisms. The first one consists of a high speed air stream blown against a rigid, sharp edge E (Fig. 4.22), located at a certain distance d exactly above the slit S. This system is aerodynamically unstable: the airstream alternates back and forth between both sides of the edge, breaking into "rotating puffs" of air called "vortices" or "eddies," which travel upward along both sides of the edge. As the velocity of the stream increases, vortices are created at an increasing rate. Since they represent a periodic perturbation of the ambient air, sound waves are generated when the vortex generation rate falls

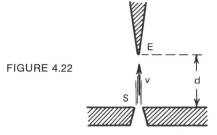

FIGURE 4.22

into the audio domain. The resulting sound is called an *edge tone*.[9] The edge tone mechanism is the primary excitation process for all wind instruments of the flute family and for flue organ pipes. The air stream oscillations are in general complex; for very small flow intensities they become nearly sinusoidal. The fundamental frequency of a free edge tone depends on the air stream velocity v and the distance to the edge d (Fig. 4.22). In the low frequency range it is proportional to the ratio v/d, i.e., it increases with increasing v and decreasing d.

The other excitation mechanism of importance in music is the *reed*, a thin plate made of cane, plastic, or metal, placed in front of a slit of nearly the same shape and of slightly smaller size than the reed (Fig. 4.23). As air is blown into the cavity from below (i.e., the pressure therein is increased), the excess air flows through the small space between the slightly lifted reed and the slit into the shallot. During this flow, the reed is drawn toward the slit.[10] This eventually interrupts the flow; the reed's own elasticity opens the slit again and the whole game starts anew. In other words, the reed starts oscillating back and forth, alternately closing (partially or totally) and opening the slit. The air moves in periodic puffs into the shallot, giving rise to a sound called *reed tone*. The fundamental frequency of a *free* reed tone depends on both the elastic properties of the reed and the excess pressure in the cavity (blowing pressure). In general, the vibratory motion of a free reed is complex, except at very small amplitudes for which it is nearly sinusoidal. Some instruments (oboe, bassoon) have *double reeds*, beating against each other. Also the lips of a brass instrument player can be considered as a (very massive) double reed system.

[9] Vortices are even formed in absence of the edge, provided the slit *S* is small enough and the velocity *v* high enough. This represents the basic physics of the *human whistle* where slit size (lip opening) and velocity of the air stream (blowing pressure) determine the fundamental frequency.

[10] By the difference in *dynamic* pressure (not static pressure) on both sides of the reed—the same effect that keeps a flying aircraft aloft!

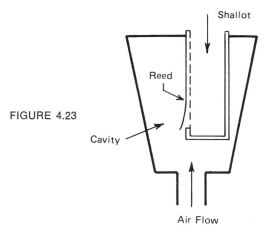

FIGURE 4.23

The edge and reed tones discussed above are seldom used alone ("free" edge and reed tones). In the woodwind instruments they merely serve as the primary excitation mechanism, the energy supplier to the air column in a pipe. In those cases not only the spectrum, but also the frequency of the vibrations of the air stream or the reed are controlled by the air column via a (nonlinear) feedback mechanism. This is accomplished by the sound waves in the air column: the first compressional wave pulse to be produced travels along 'the pipe, is reflected at the other end (open or closed), and comes back to the mouthpiece (as a rarefaction pulse in open pipes, as a compression pulse in closed ones). There it causes a pressure variation that in the case of woodwinds "overrides" all other forces (aerodynamic or elastic) and thus controls the motion of the air stream or the reed. The resulting pitch is quite different (usually much lower) from that evoked when the corresponding edge or reed tone mechanism is activated freely, in absence of the pipe. This is quite different from the case of a string mounted on a soundboard, whose pitch remains practically unaffected by the resonator. In the case of brass instruments, the mass of the player's lips is so large that the feedback from the pipe can only influence, but not override, their vibration; the latter must be controlled by the player himself by adjusting the tension of his lips. There are a few musical instruments with open reeds (accordion, harmonica, reed organ).

The tone-buildup process in a wind instrument is very complicated and not yet fully investigated. But it is of capital importance for music. In many instruments, upper harmonics build up faster than the fundamental; sometimes this can be artificially enhanced, giving a characteristic "chiff" to the resulting tone.

To understand the *steady state* sound generation in woodwinds, brasses and organ pipes, it is necessary to analyze the resonance properties of their air columns and the coupling of the latter with the primary excitation mechanism (air stream, reed, or lips). To that effect let us state the following experimentally verified facts: (1) The primary excitation mechanism sustains a periodic oscillation that is complex, of a certain fundamental frequency, and with a series of harmonics of given spectrum. (2) Fundamental frequency and spectrum of the primary oscillations are controlled (influenced, in the case of brass instruments) by the resonance properties of the air column; the total amplitude of the oscillations is determined by the primary energy supply (total air stream flow, blowing pressure). (3) The spectrum of the pressure oscillations outside the instrument (generated sound wave) is related to the internal spectrum by a transformation that is governed by the detailed form and distribution of fingerholes and/or by the shape of the bell.

To explore the resonance properties, i.e., the resonance curve, of the air column in a given wind instrument, one must devise an experimental set-up in analogy to the alternating-current-driven metal string (Section 4.1) or the vibrator-excited violin plate (Section 4.3). This is accomplished by replacing the "natural" excitation mechanism with a mechanical oscillation driver (e.g., an appropriate speaker membrane) and by measuring with a tiny microphone the pressure oscillation amplitudes in the mouthpiece (where a pressure antinode is formed in this setup). The resonance curve is then obtained by plotting the pressure oscillation amplitudes as a function of frequency, for *constant* driver oscillation amplitude. The measured amplitudes are usually expressed in decibels [R in expression (4.4)], referred to some standard level. Curves of this type are also called *input impedance* plots. Figure 4.24 sketches typical resonance curves obtained for clarinet-type and oboe-type air columns (Benade 1971) (*without* mouthpieces, bells, and open fingerholes). It is important to note that the resonance peaks obtained in this manner correspond to the vibration modes of an air column *closed* at the end of the primary driver, i.e., to a real case in which a *reed* is used as excitation mechanism at one end of the column (pressure antinode, vibration node, at the place of the reed). To find the resonance curve of the same air column corresponding to the case in which it is excited by an air stream (flutes, recorders, organ flue pipes), it is sufficient to plot the negative $-R$ of the values obtained in the previous measurement[11]: resonance peaks become dips and

---

[11] Only if R is expressed in *decibels*.

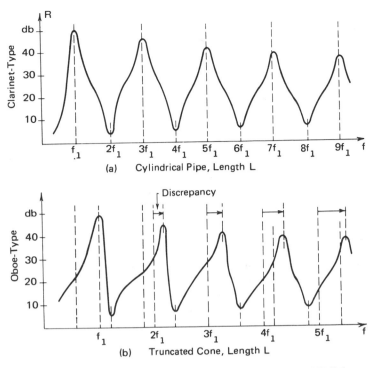

FIGURE 4.24. Typical resonance curves (after Benade 1971) for clarinet-type (cylindrical) and oboe-type (conical) air columns (without mouthpiece, bell; closed fingerholes).

dips become peaks (just turn Fig. 4.24 upside down). The main justification for this procedure is that, in the mouthpiece, the pressure antinode (that appears when a reed is used) is replaced by a vibration antinode, i.e., a pressure node, in the case of a flute with an open mouthhole.

Note in Fig. 4.24 that resonance peaks are not at all "sharp"; possible oscillation modes thus do not correspond to unique, discrete frequencies, as it appeared to be in the case of infinitely thin air columns (Section 4.4). Furthermore, in the case of the truncated cone (b), the resonance peaks are asymmetric and inharmonic (see discrepancy from the harmonic multiples $2f_1$, $3f_1$, etc.). Toward high frequencies, the resonance peaks of the truncated cone resemble those of the cylinder. On the other hand, if the cone were complete (till the tip), the resonance peaks would all lie close to the dips of curve (a) (even harmonics of the cylinder) with very little inharmonicity.[12]

Let us discuss qualitatively how the resonance curve controls the primary excitation mechanism, say, of a reed. For

[12] In real instruments, clarinet-type resonance curves (b) also display a discrepancy from harmonicity (Backus 1974).

very low intensities (small amplitudes of the reed), its motion is nearly sinusoidal, and, in principle, any resonance peak frequency (Fig. 4.24) could be evoked. In practice, however, it is found that only the frequency corresponding to the *tallest* resonance peak is excited at very low intensity level (pianissimo). Usually, this is the peak with the lowest resonance frequency; the evoked sound pertains to the *"low register"* of the instrument.

As the amplitude of the reed oscillation increases (by increasing the blowing pressure), the nonlinear character of the feedback from the air column destroys the harmonic, sinusoidal, vibration of the reed, upper harmonics appear with increasing strength (in general, the intensity of the n–th harmonic grows proportionally to the $2n$–th power of the intensity of the fundamental), and the resulting sound becomes "brighter." At the same time, the fundamental frequency readjusts itself if the upper resonance peaks are somewhat inharmonic. The rule governing this pitch readjustment is the following: the fundamental frequency locks into position in such a way as to *maximize the weighted average height of all resonance values* $R_1, R_2, R_3, \ldots$[13] corresponding to the harmonics $f_1, 2f_1, 3f_1 \ldots$ (Benade 1971). If, for instance, the upper resonance peaks deviate from harmonicity as shown in Fig. 4.24b, the pitch of the tone must become sharper as its intensity increases, in order to accommodate the set of increasingly important upper harmonics and place each of them as close as possible to a resonance peak. Due to this effect, a truncated cone will not work as a woodwind unless something is done to minimize the inharmonicity.

An interesting situation arises with the clarinet-type resonance curve (Fig. 4.24a). There, the resonance peaks are situated at only odd integer multiples of the fundamental (see also Section 4.4). Hence, all even harmonic components of the reed oscillation will be strongly attenuated. Starting from a piano-pianissimo in the low register (single-frequency excitation at the fundamental peak) and gradually increasing the blowing pressure will at first tend to evoke the second harmonic. Its energy, however, will be efficiently drained because of the dip at the corresponding frequency (Fig. 4.24a). The resulting increase in loudness (and "brightness") will thus be considerably less for a given increase in blowing pressure than in an oboe-like air column (curve *b*) where the second harmonic can build up unhindered. This is why transitions from *ppp* to *pp* are more easily managed in a clarinet than in an oboe or saxophone.

Finally, another noteworthy fact in Fig. 4.24 is the almost

---

[13] Weighted with the corresponding spectral intensity values, $I_1, I_3, I_3, \ldots$

identical location of the *dips* in both curves. When these dips are converted into peaks by plotting $-R$ instead of R to obtain the resonance curve of these air columns when they are excited with an air stream (open mouthhole), one obtains a practically identical harmonic series in both examples. Consequently, cylinder and truncated cone can be used almost indistinctly to make flute-type instruments.

So far we have considered standing waves in which the fundamental frequency is determined by the first (lowest frequency) peak of the resonance curve, yielding the low register tones of a woodwind instrument. In the "*middle register*," the fundamental frequency lies close to the second resonance peak. In reed instruments this is realized by decreasing the size of the first resonance peak below that of the second and by shifting its position away from the harmonic series. The register hole or speaker hole accomplishes this function. In the flute, this transition, or "*overblow*," is accomplished by means of a change (increase) of the air speed blown against the wedge. Notice that in the first overblow of a clarinet-type reed the pitch jumps to the third harmonic or *twelfth* (second peak, Fig. 4.24a), whereas conical-bore reeds (and all flutes) have their first overblow on the *octave* (second harmonic, Fig. 4.24b).

Another overblow leads to the *top register* of the woodwinds, with fundamental frequencies based on the third and/or fourth resonance peaks. To accomplish this with a reed instrument, the first two resonance peaks must be depressed and shifted in frequency to destroy the harmonic relationship.

Organ pipes work on essentially the same principles as a flute (flue pipes, open and stopped) or as a reed woodwind (reed pipes). The main difference is that since there is one pipe for each note for a given stop, tone holes and overblowing are unnecessary. Organ pipes are always operated in the low register (with a few exceptions in "romantic" organs). Resonance curves of open-flue organ pipes have peaks located near the integer multiples of the fundamental frequency, with a slight inharmonicity which depends on the ratio $r =$ diameter/length. Stopped pipe resonance curves resemble that of the upper graph in Fig. 4.24 with maxima at odd multiples of the fundamental. The larger the value of r, the greater will be the inharmonicity of the upper resonances. As a result, there will be a shift in the fundamental frequency of the resulting tone, plus an increasing attenuation of the higher harmonics [which will be increasingly displaced away from the (inharmonic) resonance peaks]. Consequently, *the sound of wide organ pipes is less rich in upper harmonics* ("flutey" sound). Narrow pipes (small r) have resonance peaks that lie closer to the integer multiples of the fundamental frequency, and there will be a stronger excitation of the higher harmonics (the sound is "bright" or "stringy"). The fundamental frequency is slightly misplaced with respect to the value given by relation 4.6 (open pipe) or 4.7 (stopped pipe). These relations, however, may still be used if a correction of value $0.3 \times$ diameter is added, for each open end, to the length L ("end correction"). Reed organ pipes range from the type in which the reed vibration is strongly controlled by the feedback from the air column (e.g., trumpet-family stops) to a type in which the reed vibration is practically autonomous (regal-family stops).

**4.6
Sound
Spectra of
Wind Instrument
Tones**

Resonance characteristic of the air column and excitation mechanism collaborate to determine the power spectrum and the intensity of the standing wave in the bore of the instrument. Fig. 4.24 showed two hypothetical resonance curves; real instruments, however, display a more complicated behavior due to the peculiar shape of the mouthpiece (and mouthpipe), the shape and distribution of open fingerholes, the effect of the bell, and, in the case of flutes, the effect of air speed on the width and position of the resonance peaks (Benade 1971). Here we can only summarize briefly the most important effects. The *fingerholes*, besides of course determining the effective length of the air column and, hence, the absolute position of the resonance peaks, are partly responsible for a *cutoff* of the resonance peaks above 1,500–2,000 Hz. This cutoff has an important effect on timbre (attenuation of high harmonics) and on the dynamic control of loud woodwind tones, particularly in the middle and high registers. In the oboe, the reed cavity and the constriction in the staple contribute to decrease the inharmonicity of the truncated cone resonances.

The brasses deserve special attention in this section. As already pointed out, the feedback mechanism is less efficient in the determination of the fundamental frequency, and the player must set the buzzing frequency of his lips near to the wanted frequency in order to elicit the right pitch. In a brass instrument the upper harmonics are created by the oscillating resonance properties of the mouthpiece cause by the alternate opening and closing of the lips (Backus and Hundley 1971) rather than by a feedback-controlled nonsinusoidal motion of the latter. Mouthpiece, tapered mouthpipe, main cylindrical bore, and bell combine in such a way as to yield a characteristic resonance curve rather different from that of a typical woodwind. Figure 4.25 gives an example (Benade 1971). Notice the marked cutoff frequency (mainly determined by the bell) and the large hump of peaks and dips in the mid-frequency range (mainly governed by the shape of the mouthpiece). This hump plays a crucial role in shaping the tone quality of

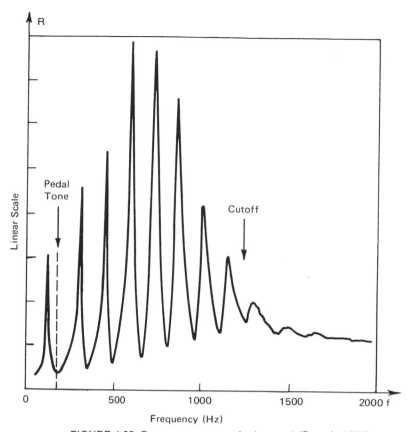

FIGURE 4.25. Resonance curve of a trumpet (Benade 1971) (given in linear scale).

By permission of Professor A. Benade, Case Western Reserve University, Cleveland, Ohio.

brass instruments. Finally, the first resonance peak lies *below* the fundamental frequency (marked by the arrow) corresponding to the rest of the peaks. Notice also the characteristic asymmetry of the peaks in the low frequency range (similar to those of a truncated cone, Fig. 4.24b), as compared to the high frequency region (where their shape is inverted).

Brasses do not have tone holes to alter the effective length of their air columns—changes in pitch are mainly effected by overblowing, i.e., making the fundamental frequency jump from one resonance peak to another. This is accomplished by appropriately adjusting the lip tension. Up to the 8th mode can be reached in the trumpet, and the 16th mode in the French horn. To obtain notes between resonance peaks, a system of valves offers a limited choice of slightly different pipe lengths. In the trombone, a continuous change tinuous change of tube length (hence of pitch) is possible through the slide. Since the lowest resonance peak is out of tune with the rest of the almost harmonic series of peaks, it cannot be used. Rather, the fundamental frequency of the lip vibration is set at the value of the *missing* fundamental (arrow in Fig. 4.25) that corresponds to the second, third, etc. peaks. This leads to the so-called *"pedal note"* of a brass instrument (used only in the trombone). It can be played only at considerable loudness levels.

The spectral composition of the sound waves emitted by a wind instrument is different from that of the standing vibrations sustained in its air column. The bell and/or the open fingerholes are mainly responsible for this spectral transformation. This transformation leaves the spectral composition above the cutoff frequency relatively unchanged, while it tends to attenuate the lower harmonics. In other words, woodwind and brass tone spectra are richer in higher harmonics than the vibrations actually produced inside the instrument. For a recent review of the physics of brass instruments, see Benade (1973).

Some wind instrument tone spectra have formants, i.e., characteristics that are independent of the fundamental frequency of the tone (Section 4.3). The bassoon and English horn are examples, with (not too well-defined) spectral enhancement around 450 and 1,100 Hz, respectively. These formants, however, are caused by the excitation spectrum characteristics of the *double reeds;* they are not determined by the resonance properties of the instrument's bore. Although not an explicit topic in this book, we must mention here the *human voice* as the most notable example of a "wind instrument" in which formants play a crucial role: they are the determining characteristic of all *vowel sounds.* Formants in the human voice are mainly determined by the resonance properties of the nasopharyngeal cavity (Flanagan 1972). The shape of this cavity determines which of two main frequency ranges of the vocal chord vibrations are to be enhanced. These, in turn, determine whether the outcoming sound is "ah," "ee," "oh," etc.

4.7 Trapping and Absorption of Sound Waves in a Closed Environment

Musical instruments are usually played in rooms, halls, auditoriums, and churches. The sound a listener perceives under these conditions is not at all identical to the one emitted by the instrument. For this reason, the enclosure in which an instrument is played may be considered a natural extension of the latter, with the difference that whereas a given musical instrument has certain immutable acoustical properties, those of the enclosure may vary widely from case to case and from place to place. The subject of auditorium acoustics is as important for music as is the physics of musical instruments.

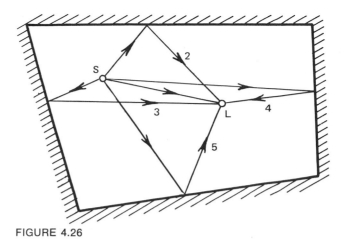

FIGURE 4.26

To analyze the effect of an enclosure on a musical sound source placed somewhere inside, let us consider a musical instrument at position S and a listener at position L (which may coincide with S if the listener is the player) in a room of perfectly reflecting walls (Fig. 4.26). The instrument starts playing a given note at time $t=0$, holding its intensity steady afterwards. We assume that the sound is emitted equally into all directions. As sound waves propagate away from S, the listener will receive a first signal after the short interval of time SL/V which it takes the *direct* sound to travel from S to L [for instance, if $SL = 10$ m, $V = 334$ m/sec [relation (3.6)], the time of direct arrival is 0.03 sec; for the player it is practically zero.] As we shall see in Section 5.1, the direct sound plays a key role in the perceptual process (*precedence effect*). Immediately thereafter, reflected waves (trajectories 2, 3, 4, 5, etc.) will pass through point L in rapid succession (in the figure, the reflections on the floor and the ceiling have been ignored). The first few reflections, if very pronounced and well separated from each other, are called echoes. This game continues with secondary, tertiary, and multiple reflections (not shown in the figure). As time goes on, and the instrument keeps sounding, acoustical energy passing through point L will keep building up. If there were no absorption at all, the sound waves would "fill" the room traveling in all directions and the acoustical energy emitted by the instrument would accumulate and remain trapped in the enclosure; the loudness would thus build up gradually at any point inside.[14]

[14] For almost any kind of shapes of rooms and positions of the source therein, there may be regions practically inaccessible to sound waves emitted from S ("blind" spots), or regions into which sound waves are focused (e.g., the focal points in elliptic enclosures).

In the real case, of course, there is absorption every time a sound wave is reflected. Hence, the sound wave intensity will not increase indefinitely but level off when the power dissipated in the absorption processes has become equal to the rate at which energy is fed in by the source (a similar situation to the vibration build-up in a bowed string, Fig. 4.10). This "equilibrium" intensity level I_m of the diffuse sound is much higher than that of the direct sound (except in the neighborhood of the sound source).

When the sound source is shut off, an inverse process develops: first, the direct sound disappears, then the first, second, etc., reflections. Figure 4.27 schematically depicts the behavior of the sound intensity at a given point in a typical enclosure. The sound decay, after the source has been shut off, is called *reverberation* and represents an effect of greatest importance in room acoustics. This decay is nearly exponential (e.g., see Fig. 4.8); quite arbitrarily one defines *reverberation time* as the interval it takes the sound level to decrease by 60 db. According to Table 3.2 this represents an intensity decrease by a factor of 1 million. Desirable reverberation times in good medium-sized concert halls are of the order of 1.5–2 seconds. Longer times would blur too much typical tone successions; shorter times would make the music sound "dry" and dull (see Section 4.8).

We may discuss a few simple mathematical relationships that appear in room acoustics. Let us imagine an enclosure of perfectly reflecting walls with no absorption whatsoever, but with a built-in *hole* of area A. Whenever the maximum intensity I_m is reached (Fig. 4.27), acoustical energy will be escaping through the hole at a rate given by the product

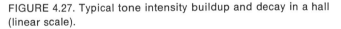

FIGURE 4.27. Typical tone intensity buildup and decay in a hall (linear scale).

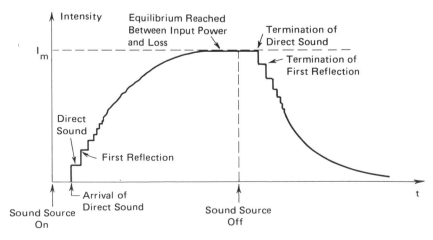

I_mA.[15] Since this corresponds to the steady state in which the power P supplied by the instrument equals the energy loss rate, we can set $P = I_mA$, or

$$I_m = \frac{P}{A} \tag{4.9}$$

In a real case, of course, we do not have perfectly reflecting walls with holes in them. However, we still may *imagine* a real absorbing wall as if it were made of a perfectly reflecting material with holes in it, the latter representing a fraction a of its total surface. a is called the *absorption coefficient* of the wall's material. A surface of S square meters, of absorption coefficient a, has the same absorption properties as a perfectly reflecting wall of the same size but with a hole of area $A = Sa$. Absorption coefficients depend on the frequency of the sound (usually increasing for higher frequencies), and have values that range from 0.01 (marble; an almost perfect reflector) to as much as 0.9 (acoustical tiles). Taking all this into account, we can rewrite relation (4.9) in terms of the actual wall surfaces S_1, S_2, . . . with corresponding absorption coefficients a_1, a_2, . . . :

$$I_m = \frac{P}{S_1a_1 + S_2a_2 + \cdots} \tag{4.10}$$

This relation can be used to estimate auditorium sizes needed to achieve wanted values of I_m, for a given instrument power P, and a given distribution of absorbing wall materials.

The reverberation time τ_r is found to be proportional to the volume V of the hall and inversely proportional to the absorbing area of the walls $A = S_1a_1 + S_2a_2 + \ldots$. Experiments show that, approximately,

$$\tau_r = 0.16 \frac{V}{S_1a_1 + S_2a_2 + \cdots} \tag{4.11}$$

with V in cubic meters, S in square meters, and τ_r in seconds. Since the absorption coefficients usually increase with sound frequency, τ_r will decrease with increasing pitch: bass notes reverberate longer than treble notes.

One of the problems in room acoustics is that the *audience* influences greatly (increases) the absorption properties of a hall. This must be taken into account in the design of auditoriums. In order to minimize the effects related to the unpredictable size of an audience and its spatial distribution, it would be necessary to build a seat whose absorption coefficient is nearly independent of whether it is occupied

---

[15] It is assumed here that I_m represents the diffuse, *omnidirectional* sound energy flow.

or not. The absorbing effect of the audience is maximally felt in enclosures with very long reverberation times, such as in churches and cathedrals. No single performer is so exposed to (and harassed by) a changing acoustical environment as an organist.

Tone distribution, build-up and decay, as well as the frequency-dependence of the absorption coefficients, have a profound effect on the physical characteristics of musical tones emitted by an instrument in a real environment, hence on the perception of music by the listener. The time dependence of tones is deeply affected: transient characteristics are altered, and, for instance, a staccato note becomes stretched in time depending on the reverberation properties of the hall. The tone spectrum is also affected, since the absorption coefficients are all frequency dependent. Finally, taking into account that the phases of the waves passing through a given point in a field of reverberant sound are randomly mixed, it can be shown that the resulting SPL of each harmonic component will also fluctuate at random, introducing a limit to the listener's ability to recognize timbre in a closed musical environment (Plomp and Steeneken 1973).

There are other second-order effects, usually neglected, related to a wave phenomenon called *diffraction*. When a sound wave hits an obstacle (e.g., a pillar in a church or a person sitting in front of the listener) three situations may arise: (1) If the wavelength of the sound wave is much smaller than the size (diameter) of the obstacle (e.g., a high-pitch tone) (Fig. 4.28a), a sound "shadow" will be formed behind the obstacle, with "normal" reflection occurring on the front side. (2) If obstacle and wavelength are of roughly the same magnitude, a more complicated situation arises, in which the obstacle itself acts as a sound reemitter, radiating into all directions (not shown in Fig. 4.28). (3) If the wavelength is much larger than the obstacle (Fig. 4.28b) (e.g., bass tones), the latter will not disturb the sound wave at all, which will propagate almost undisturbed. Regular arrangements of obstacles (as the distribution of seats or people in the audience) may lead to inter-

FIGURE 4.28

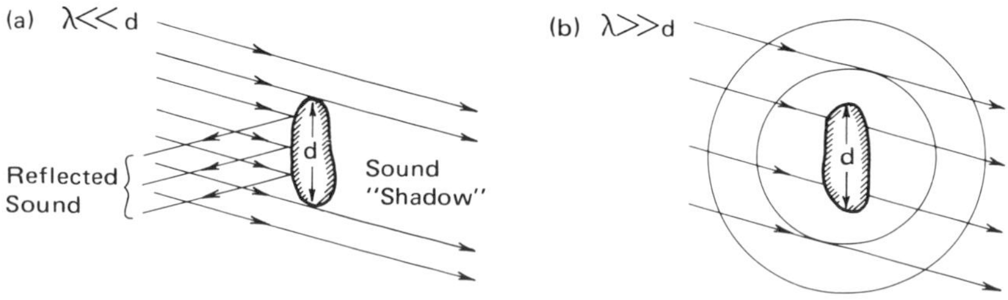

ference patterns for given wavelengths and directions of propagation. Finally, standing waves may build up for certain configurations of the enclosure, certain frequencies, and certain positions of the source. This leads to the formation of annoying nodes and antinodes (Section 3.3) in the room.

Diffraction and the precedence effect play an important role in electroacoustical sound reproduction. Electronic compensation for sound wave diffraction effects at the head of the listener is desirable when stereophonic signals are received from two loudspeakers (Damaske 1971). A correct stereo reproduction of the "direct sound" (i.e., the precedence effect) is necessary to prevent headphones from giving the sensation of a sound image localized "inside the head" (Laws 1973; see also pp. 40 and 53).

**4.8
Perception of
Pitch and
Timbre of
Musical Tones**

Whereas considerable research has been done on the perception of pitch and loudness of pure tones (Sections 2.3, 2.9; 3.4, 3.5), much remains to be done in the study of the perception of complex tones. That the timbre of a tone can be modified by reinforcing certain overtones has been known for many centuries. Indeed, genuine *tone synthesis* was first performed by pipe organ builders in the thirteenth or fourteenth century. Organs of those times did not have multiple stops; rather, each key sounded a fixed number of pipes called "Blockwerk," composed of one or several pipes tuned to the fundamental pitch of the written note, plus a series of pipes respectively tuned to the octave, the twelfth, the fifteenth, etc., following the series of upper harmonics (excluding the seventh). The particular combination of loudness chosen for each component pipe determined the particular quality of sound of the instrument. Later on, the first multiple stops appeared: they allowed the organist to selectively turn on or off the various ranks of pipes corresponding to the upper harmonics in the "Blockwerk," and thus choose from among several options the particular timbre of sound of the organ (and alter the loudness—Section 3.4). It was only one or two centuries later that new independent stops were added, in the form of ranks of pipes of individually different timbre.

Synthesis of sound is thus rather old hat. However, *analysis* of sound, i.e., the individualization of upper harmonics that appear simultaneously in a naturally produced tone, was not explicitly mentioned in the literature until 1636, when the remarkable French scientist-philosopher-musician, Père Mersenne published the first study of the (qualitative) analysis of the upper harmonics present in a complex tone.

Two main questions arise regarding perception of complex tones: (1) why does a complex tone, made up of a superposition of different frequencies, give rise to only *one* pitch sensation; (2) what is it that enables us to distinguish one tone spectrum from another, even if pitch and loudness are the same? Although we have already answered in part the first of the preceding questions (Section 2.9), it is useful to reexamine once more the perception process of a *complex* sound wave impinging on the eardrum. The eardrum will move in and out periodically with a vibration pattern dictated by the complex, nonsinusoidal vibration pattern of the wave. This motion is transmitted mechanically by the chain of ossicles to the oval window membrane, which reproduces nearly the same complex vibration pattern. Neither eardrum nor bone chain "knows" that the vibration they are transmitting is made up of a superposition of different harmonics. This analysis is only made in the next step.

The complex vibration of the oval window membrane triggers traveling waves in the cochlear fluid. This is the stage at which the separation into different frequency components takes place. As stated in Section 2.3, the resonance region for a given frequency component (region of the basilar membrane where the traveling wave causes maximum excitation) is located at a position that depends on frequency. A complex tone will therefore give rise to a whole multiplicity of resonance regions (Fig. 2.24), one for each harmonic, the actual positions of which can be obtained using Fig. 2.8 as a guide. In view of the near-logarithmic relationship between x and f, the resonance regions will crowd closer and closer together as one moves up the harmonic series (Fig. 2.24). Because each resonance region is not "sharp" but extended over a certain length (Section 2.4), overlap between neighboring resonance regions will occur, particularly for higher harmonics. Actually, beyond about the seventh harmonic, all resonance regions overlap, lying within a critical band, and it makes little sense to consider them individually.[16] Each one of the resonance regions of the basilar membrane oscillates with its own resonance frequency and with a phase that is related (but not equal) to the phase of the corresponding harmonic (Fourier component) in the original eardrum oscillation (e.g., Flanagan 1972). Overlapping, of course, complicates this picture considerably.

A single complex tone thus elicits an extremely complicated situation in the cochlea. Why, then, do we really perceive this tone as *one* entity, of well-defined pitch, loudness, and timbre? As explained in Section 2.9, this is belived to be the result of a *spatial pattern recognition pro-*

[16] Experiments have indeed shown that for usual musical instruments, due to this overlap (frequency discrimination, Section 2.4), harmonics beyond about the seventh cannot be heard out (Plomp 1964).

cess, i.e., an auditory *Gestalt* perception. The characteristic feature that is recognized in this process, that is common to all periodic tones regardless of their fundamental frequency and Fourier spectrum, is *the nearly invariant distance relationship between resonance maxima* on the basilar membrane. The pitch sensation is to be regarded as the "final output signal" of this recognition process.[17] The crucial tone components that activate the central pitch processor responsible for this recognition process are the first six to eight harmonics. Higher harmonic resonances overlap on the basilar membrane, and the distance relationship among them becomes "fuzzy" (Section 2.9). The recognition mechanism of the pitch processor (the "template fitting," the "spatial autocorrelation," or the "learning matrix" discussed in Section 2.9) can work successfully even if part of the input is missing (e.g., a suppressed fundamental). In such a case it may commit matching errors and yield ambiguous or multiple pitch sensations. A more detailed discussion of how this recognition process may actually work is given in *Appendix II*.

We should point out once more that neural pulse time-distribution analysis (Section 2.9) cannot be ruled out completely as a possible mechanism cooperating in the determination of the subjective pitch of a complex tone. We have almost forgotten the other remarkable invariant feature of the excitation pattern elicited by a complex tone (Section 2.7) *pair of consecutive components of a harmonic series yield one common repetition rate* (Fig. 2.19). Although there is less overlap among lower order harmonics, the repetition rate is most clearly defined between them (Section 2.7), and all primary neurons wired to the relevant region should fire pulse sequences carrying information on this repetition rate (Fig. 2.22). As a matter of fact, as stated on p. 57, psycho-acoustical experiments with very low frequency tones of very short duration seem to favor a time-cue process (only as few as 2–3 cycles can give rise to a clear pitch sensation—a fact that cannot be explained satisfactorily by a spectral analysis theory).

Pitch perception of a complex tone is just one output from among a whole sequence of cognitive tasks. It is the psychophysical mechanism that transforms the peripheral activity pattern evoked by a musical tone or a voiced speech element, into another pattern in such a way that all stimuli with the same periodicity are similarly represented—yielding the same pitch sensation. An optical analog would be the recognition of a given letter—irrespective of its orientation, size, color, or type (Section 2.9). Whenever our eardrum is set into a periodic vibration, however complex, the

---

[17] We strongly recommend that the reader review Section 2.9.

fact that we do perceive *one* pitch tells us that our nervous system got the message: "aha, a periodic tone" (in contradistinction to a noise, or to the multiple, ambiguous pitches that one may perceive from *in*harmonic tones). Just as when we look at either one of the symbols A a ∀ a our brain gets the message: "aha, the letter A."

But then, of course, there are other features of the primary auditory stimulus (disregarded by the pitch processor) that yield perceptional output from other stages of the sound pattern recognition process. In the visual example, for instance, we also perceive whether a given letter is big or small. Likewise, for a complex tone, *the perception of loudness is linked to the total rate of neural pulses* reaching the upper stages of the auditory nervous system (Sections 2.9 and 3.5).

Another most important recognizable feature of a complex tone yields the sensation of *tone quality* or *timbre*. Here, we must make a clear distinction between the static situation that arises when we listen to a steady-sounding complex tone of constant fundamental frequency, intensity, and spectrum, and the more realistic dynamic situation when a complex tone with transient characteristics is perceived as part of a musically relevant context. Let us first analyze the static case. Psychoacoustical experiments with electronically generated steady complex tones, of equal pitch and loudness but different spectra and phase relationships among the harmonics, show that the timbre sensation is controlled primarily by the power spectrum (Plomp 1970). Phase changes, although clearly perceptible, particularly when effected among the high frequency components, play only a secondary role. The *static* sensation of quality or *timbre thus emerges as the perceptual correlate of the activity distribution evoked along the basilar membrane—provided that the correct distance relationship among resonance peaks is present* to bind everything into a "single-tone" sensation. By dividing the audible frequency range into bands of about one-third octave each (roughly corresponding to a critical band, Section 2.4), and by measuring the intensity or sound energy flow that for a given complex tone is contained in each band, it was possible to define quantitative "dissimilarity indices" for the (steady) sounds of various musical instruments, which correlate well with psychophysically determined timbre similarity and dissimilarity judgments (Plomp and Steeneken 1971). It is important to point out that the timbre sensation is controlled by the *absolute* distribution of sound energy in the critical bands, not by the intensity values relative to that of the fundamental. This is easily verified by listening to a record or magnetic tape played at the wrong speed. This procedure leaves relative spectra unchanged, merely shifting all frequencies up or down; yet a clear change in timbre of all instruments is perceived.

The static timbre sensation is a "multidimensional" psychological magnitude related not to one but to a whole set of physical parameters of the original acoustical stimulus (the set of intensities in all critical bands).[18] This is the main reason semantic descriptions of tone quality are more difficult to make than those of the "unidimensional" pitch and loudness. Except for the broad denominations ranging from "dull" or "stuffy" (few upper harmonics), to "nasal" (mainly odd harmonics), to "bright" or "sharp" (many enhanced upper harmonics), most of the qualifications given by musicians invoke a comparison with instrumental tones ("flutey," "stringy," "reedy," "brassy," "organ-tone-like," etc.).

Here we come to the more realistic dynamic situation. Our subjective reaction to complex tones depends quite appreciably upon the *context* of which they are a part (Houtsma and Goldstein 1972). The performance of tasks that are musically "meaningful," such as the identification of the tone source, i.e., the instrument, and the recognition of melodies or harmonies greatly influences how complex tones are processed in the brain. This applies even to the pitch perception: Experiments with electronically generated spectral components convincingly showed that the attachment of a single pitch to complex sounds is greatly facilitated by, or sometimes even requires, the presentation of the test tones in the form of a "meaningful" melody[19] (Houtsma and Goldstein 1972). Individual electronically synthesized complex tones, taken out of a musical context, often may lead to ambiguous or multiple pitch sensations.

When listening to a complex tone, *our auditory system pays more attention to the output of the central pitch mechanism* (which yields *one* unique pitch sensation) *than to the primary pitch of the individual harmonic components.* If we want to "hear out" the first six or seven upper harmonics of

[18] Although there are about 15 critical bands in the musically relevant frequency range the intensities of which ought to be specified in order to determine the spectrum, a study of vowel identification (Klein, Plomp, and Pols 1970) indicates that only *four* independent intensity parameters (each one a specific linear combination of the intensities in all critical bands) are sufficient to specify a complex tone within the "timbre resolution capability" of the auditory system.

[19] A startling experiment to test this context-dependent fundamental pitch tracking effect can be performed on the organ. Play a piece (e.g., Bach's Orgelbüchlein chorale No. 42) with a single soprano melody on a Cornet-type combination $8' + 4' + 2\frac{2}{3}' + 2' + 1\frac{3}{5}' + 1\frac{1}{3}' + 1'$, accompanied at *all* times with a soft $8' + 4'$ and $16' + 8'$, respectively. Ask a musically trained audience to carefully monitor the pitch of the melody, but warn them that there will be changes of timbre. After the first 5-6 bars, repeat the piece, but eliminate the 8' from the melody. Repeat again, eliminating the 4'; then the 2', finally the 1'. At the end, make the audience aware of what was left in the upper voice and point out that the pitch of the written note was absent altogether (in any of its octaves)—they will find it hard to believe! A repetition of the experiment, however, is likely to fail—because the audience will re-direct their pitch-processing strategies!

a steadily sounding complex tone, we must command a "turn-off" (inhibition) of the dominating periodicity pitch mechanism and focus our attention on the initially choked output from the more primitive primary or spectral pitch mechanism, determined by the spatial position of the activated regions of the basilar membrane (Section 2.3). This inhibiting and refocusing takes a considerable time—considerably longer than the buildup of the general tone-processing mechanism (Section 3.5). This is why upper harmonics cannot be "heard out" in short tones or rapidly decaying tones.[20] It is important to remark that, in view of the effect discussed briefly on page 91, primary pitch matching of the overtones of a complex tone *always yields slightly "stretched" intervals*, for instance, between the first and the second harmonic (stretched octave), and so on (Terhardt, 1971). This shift is caused by the perturbing influence of the ensemble of all other harmonics on the one harmonic whose primary pitch is being matched. The effect is small (up to a few percent), but may be musically relevant (Sections 5.4 and 5.5).

4.9
Identification of
Musical Sounds

A characteristic feature of all pattern recognition processes is the "selective information loss." This is related to the nervous system's minimum effort–maximum efficiency way of operating: In order to be able to sort out *meaningful* information-carrying stimuli from the awesome background of total sensorial input (and thus be able to identify objects and their causal interrelationships) the system must rely on a series of "filters" that help screen irrelevant input features from the relevant ones. These "filters" must be "tuned" to certain invariant signatures of the stimuli that are considered "relevant" (by genetic information transfer, or as the result of a learning process). In sound perception, the first, most "primitive," input feature to be recognized by the nervous system is probably intensity, with the perceptual correlate *loudness* This correlate does not depend much on any other fine structure of the stimulus except total acoustic energy flow (Section 3.5). The next auditory input feature to be considered is the periodicity of a sound (as represented by the spatial distribution of resonance maxima, or by the temporal distribution of neural pulses). From this recognition process, the sensation of subjective *pitch* is extracted. The third level of refinement is the consideration of the power spectrum of the tone, leading to the sensation of *timbre*.

Timbre perception is, however, just a first stage of the operation of *tone source recognition*—in music, the identi-

[20] The fact that the seventh harmonic is a dissonance has been worrying musicians for a long time. This worry is unfounded though. The seventh harmonic is extremely difficult to be singled out, even in steady, electronically generated complex tones (Plomp 1964).

fication of the instrument. From this point of view, tone quality perception is the mechanism by means of which information is extracted from the auditory signal in such a way as to make it suitable for: (1) *storage* in the memory *with an adequate label of identification,* and (2) *comparison* with previously stored and identified information. The first operation involves *learning* or conditioning. A child who learns to recognize a given musical instrument is presented repeatedly with a melody played on that instrument and told: "This is a clarinet." His brain extracts suitable information from the succession of auditory stimuli, labels this information with the qualification "clarinet" and stores it in the memory. The second operation represents the conditioned response to a learned pattern: When the child hears a clarinet play after the learning experience, his brain compares the information extracted from the incoming signal (i.e., the timbre) with stored cues, and, if a successful match is found, conveys the response: "a clarinet." On the other hand, if we listen to a "new" sound, e.g., a series of tones concocted with an electronic synthesizer, our information-extracting system will feed the cues into the matching mechanism, which will then try desperately to compare the input with previously stored information. If this matching process is unsuccessful, a new storage "file" will eventually be opened up for this new, now identified, sound quality. If the process is only partly successful, we react with such judgments as "almost like a clarinet" or "like a barking trombone."

What are the neural processes that lead to timbre perception and tone source identification? Few details are known in this field, and we can only sketch some of the current ideas. First, we note that experiments have shown that neurons in the cortical areas to which the afferent transmission system from a sensorial organ is wired, are *"feature-detectors,"* responding to well-defined but complex physical features of the original sensorial stimulus. In the visual cortex, for instance, neurons have been found that respond only to dark or light bars or edges in a given part of the visual field, inclined with a certain angle, or to a line moving in a certain fashion (Hubel 1971). In the auditory cortex, too, there are neurons that seem to respond only to certain types of complex sounds (Simmons 1970). This exogenous, externally stimulated activity of a cortical neuron is the result of neural signal processing in the afferent pathway from the sensor to the cortex, controlled by the particular way the neurons are wired and interact with each other. A static complex sound of given harmonic spectrum probably elicits a well-defined firing pattern in a well-defined distribution of cortical neurons, and we report a given sensation of timbre. We must point out that this is very likely a *collec-*

tive phenomenon involving large ensembles of neurons; there is experimental evidence showing that, in general, one isolated neuron responds to many different kinds of input characteristics (Gerstein and Kiang 1964), thus cooperating in the detection of a large variety of input features. It is the distribution in time and space of neural activity that seems to be locked unambiguously to a given stimulus feature (in our case, a given musical sound). In more physical terms, neural response in the brain cortex represents a "holographic" image of the stimuli (e.g., Pribram 1971), in which information on one "point" (or feature) of the input is spread in space and time over a large number of "points" (or neurons) of the cortex.

When neural activity is displayed on the cortex, we report a given sensation. However, experiments have shown that *the cortical activity evoked by a given stimulus is profoundly altered if the information carried by that stimulus has a certain meaning* (John 1972). To the peripherally triggered firing pattern, an endogenous activity or *readout signal* is added (with a typical delay of 50–80 milliseconds), triggered internally in the brain by higher order operations of comparison and identification, somehow representing the conscient experience related to the act of perception. This readout signal is found as a common, coherent pattern of activity spread over many different areas of the brain. It is absent altogether if the original signal is meaningless (i.e., not related to previously stored or learned messages) or if no attention[21] is being paid to the stimulus. Experiments with animals have shown that a readout signal may sometimes be in error, i.e., may not correspond to the externally evoked activity—in those cases the behavioral conditioned response also comes out in error!

It is believed that the replay of a given endogenous readout pattern represents the elementary act of *remembering*. When this pattern is triggered externally while we listen to a tone, we "remember" that this tone comes, say, from a clarinet. When this activity is released internally (by some association or by a volitive command), we are remembering the sound of a clarinet in *absence* of a true external sound. This then represents the most simple form of activation of the "acoustical imaging mechanism." Experiments with vision have shown that, for instance, the mere imagination of a geometrical form evokes activity in the visual cortex very similar to that externally evoked when the subject actually sees that form (Herrington and Schneidau 1968). A similar effect is likely to occur with the auditory image of musical tones and musical forms.

---

[21] "Attention" probably means to let a given readout pattern of neural activity spread over a large area of the brain by willfully inhibiting all other potential sources of readout patterns.

Nothing is known as yet on how the readout activity is actually produced. Obviously, it must involve a complex interaction between different centers of the brain. When we listen to a "new" musical instrument and at the same time are told what the instrument is, the primary exogenous auditory cortical activity evoked by the musical sound and the activity in other centers elicited by the awareness of the name or physical appearance of the instrument, interact somehow,[22] giving rise to the formation of a readout pattern that is a signature of that particular experience. During the initial phase of learning, this readout pattern can be triggered only by the *simultaneous concurrence* of the different inputs that are part of the learning experience (e.g., the instrument's sound and the instrument's appearance). As a result of repetitive stimulation, however, some long lasting alterations occur in the intervening neural tissue (probably in the form of changes in synaptic connections between the activated neurons). These alterations are such that, in the future, the same readout patterns can be elicited even if *only one* of the originally concurrent input forms appears (Roederer 1974). Thus, if we hear tones from a "known" instrument, the activation of the corresponding readout pattern tells us what instrument it is. On the other hand, the same readout pattern can be triggered from the "other end" by "thinking" about the instrument—and the spread of this activity onto the auditory cortex will allow us to "hear" its tones internally.

It is believed that the alterations appear in physically well-defined and localized pathways of enhanced neural connectivity, called neural *engrams* (Eccles 1970). This capacity of "self-organization" is believed to represent the most fundamental property of the cortical neural networks. Experiments conducted to study the effects of ablations of cerebral tissue on memory and on recognition operations point against the existence of a highly localized "seat" of memory (John 1972). Current thought is that temporal relations between firing patterns averaged over a large ensemble of brain cells are more significant than a localized spatial distribution of this activity and the actual firing pattern of a reduced number of specific neurons. Learning, according to this view, is to increase the probability of recurrence of a given coherent firing pattern in the neural network—that one triggered by the particular constellation of multiple input repeatedly experienced during the learning process. In other words, learning is not represented so much by an increase of the storage of information per se, but by appropriate modifications of the information-processing *mechanisms*, and memory emerges in this picture as the storage of information-processing *in-*

---

[22] Probably in the association areas of the cortex, where neural signals from different receiving areas converge, either through direct lines or through subcortical "switchboard" stations.

structions, rather than as the storage of information on facts (images).[23]

We have presented the perception of tone quality as being the first step of an identification mechanism—that of the instrument to which the tone belongs. Musicians will protest and say that there is far more to the sensation of timbre than merely providing cues to find out "what is playing." For instance, what is it that makes one instrument sound more "beautiful" than another of the same kind? First we should point out that this is obviously related to a further degree of sophistication of the identification mechanism mentioned above—we can learn to extract an increasingly refined amount of information from the sound vibration patterns of an instrument, so as to be able to distinguish among different samples of instruments of the same kind. *Why* some vibration patterns appear to be more "beautiful" than others is not known. A great deal of research has been attempted, for instance, to find out what physical character-istics make a Stradivarius violin a great instrument (e.g., Saunders 1946). Many of these characteristics are dynamic in character, and most of them seem to be more related to the major or minor facility with which the *player* can control the wanted tone "color" (spectrum and transients), than to a "passive" effect upon a listener. For instance, a significant aspect of violin tones seems to be related to the effect of the narrowly spaced resonance peaks (e.g., Fig. 4.15) on loud-ness and timbre when the fundamental frequency of the tone is modulated by the player in a *vibrato* (Matthews and Kohut 1973). Under such circumstances, the frequencies of the harmonic components sweep back and forth past the narrow and unequally spaced resonance peaks. As a result, the amplification of each component varies periodically, and so will loudness and timbre of the tone. Depending on the particular microstructure of the resonance curve of his in-strument, *the string player has the possibility of inducing extremely fine changes in loudness and timbre coupled to his vibrato.* To a great extent the impression on the listener is also based on learned experience: we have been *told* since the inception of our musical training which tones are to be labeled as beautiful, and which not! And our brain has learned to build up the corresponding identifying readout patterns, based on a tremendously refined information pro-cessing in order to recognize minute, practically unmeasur-able, fine strucures of the incoming acoustical signal.

[23] Computers and calculators operate in a rather similar fashion: to execute a multiplication the computer does not consult a huge multi-plication table stored in its memory—it performs the operation every time it is needed. *What* is stored in its memory are just a few *instruc-tions* on how to multiply! In the brain what is stored are *instructions on how to construct images,* not the images per se.

Superposition and Successions of Complex Tones and the Perception of Music

In the course of Chapters 2, 3, and 4 we have been moving gradually up the ladder of neural processing of acoustical signals, from the mechanisms leading to the perception of primary pitch, loudness, subjective pitch, and timbre, and to the recognition of a musical instrument. On the physical side, we have analyzed how the sound characteristics leading to these sensations are actually generated in musical instruments. These psychological attributes are necessary, but by no means sufficient, ingredients of music. Music is made of *successions* and *superpositions* of tones that convey certain meanings (Gestalt) which can be analyzed, stored, and compared in the brain. An objective study of the neuropsychology of these higher order processes is practically nonexistent. The reader no doubt already has sensed how much guesswork was involved in the discussion of the perception of timbre (Section 4.8). We are in much worse shape when it comes to the higher psychological attributes such as consonance, dissonance, the sense of tonality and return, and, in the end, the whole gamut of emotional reactions to music.

5.1
Superposition of Complex Tones

Polyphonic music consists of superpositions of complex tones. Even if only a single melody is played in monophonic music, a superposition of reverberant sounds usually reaches our ear, leading to the superposition of complex tones. The psychophysical study of complex tone superposition effects is still very much incomplete. This applies particularly to the understanding of how the ear and the brain are able to disentangle the "mess" of frequencies that belong to different simultaneously sounding complex tones, so as to keep the sensations of these tones apart. Here, we can only present a qualitative and oversimplified discussion.

When two complex tones of different pitch are superposed, either of two situations may arise: the fundamental frequency of the higher tone is equal to one of the upper harmonics of the lower tone, or it is not. In the first case, the upper tone will reinforce certain upper harmonics of the lower tone. Why don't we, then, simply detect a change in *timbre* of the lower tone, instead of clearly singling out the upper tone and even keeping the timbre of both tones apart? A similar problem arises with the second case where each tone produces its own multiplicity of resonance regions on the basilar membrane. How does our brain single out from the resulting mixture which sequence belongs to what tone? For instance, consider the superposition of two complex tones, say, A_2, (110 Hz) and $C_3\sharp$ (140 Hz). The harmonics of both tones are shown in Fig. 5.1 on a linear frequency scale. For each one of these frequencies there is a corresponding resonance region on the basilar membrane (Fig. 2.8). The sensor cells do not get the slightest cue as to which tone each resonance region belongs to. This discrimination must therefore be performed at a higher center in the auditory neural pathway.

The *pitch* of the two tones is discriminated by the central pitch processor (Section 2.9). Either "template matching," "spatial autocorrelation," or the "learning matrix," would explain the appearance of *two* prominent output signals from the processor, corresponding to the two pitches heard in the case of a double complex tone signals as shown in Fig. 5.1 (see *Appendix II*).

A most astounding capability of the auditory neural system is that of discriminating the *timbre* of two simultaneously sounding complex tones. No real music would be possible without such capability. For instance, assume that you listen monaurally with earphones to the sound of instrument 1 playing *exactly* the note A_4, and instrument 2 playing *exactly* A_5, at nearly the same intensity level. Figure 5.2 shows the hypothetical superposition. The total length of the vertical bars represents the total intensity of each harmonic actually reaching the ear. How does our brain manage to keep both tones apart, in terms of timbre? This discrimination mechanism is not at all understood; a *time element* seems to play the key role. First, the initiation (attack) of two "simultaneous" tones is never exactly syn-

FIGURE 5.1

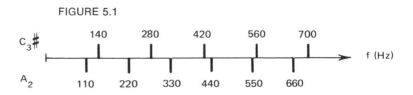

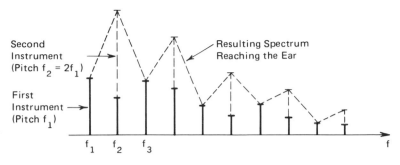

FIGURE 5.2

chronized, nor does the tone buildup develop in the same manner, particularly if both tones come from different directions (stereo effect). During this transient period, the processing mechanism in our brain seems able to lock in on certain characteristic features of each instrument's vibration pattern and to keep track of these features, even if they are garbled and blurred by the signal from the other instrument. During performance, too, slight excursions in frequency and intensity (the so-called *chorus* effect) that are *coherent* for the whole harmonic series of each tone are used by the auditory processing mechanism to trigger the right neural readout signals for each instrument (Section 4.9). Other, more pronounced, variations that seem to provide important cues to the mechanism for discriminating tone quality are the periodic variations in pitch (vibrato) and intensity (tremolo, vibrato) that can be evoked (voluntarily or involuntarily) in otherwise steady tones of many musical instruments. Superposition of multiple complex sounds totally deprived of these small coherent time-dependent perturbations—as happens when multiple stops are combined in organ music—are indeed much more difficult to discriminate in timbre.

What probably aids the discrimination mechanism of both pitch and timbre of complex tones the most is the information received from a *progression* of tone superpositions. In such a case not only the above mentioned "primary" cues given by the coherent fluctuations in timing and frequency of each tone can be used, but the "higher order" or secondary information extracted from the melodic lines (the musical "messages") played by each instrument (see also p. 137) will also be available.

The complex tone discrimination mechanism has its equivalent (and probably, its origin) in the mechanism that steers our auditory perception when we follow the speech of one given person among many different conversations conducted simultaneously at similar sound levels.[1] This ability has been pointedly called the "cocktail party effect," and very likely uses the same cues, primary and secondary, as the complex tone discrimination mechanism. Finally, it probably is this same mechanism which enables us to disentangle individual sounds from among the messy tone superposition in a reverberating hall (Section 4.7). In this latter case, it is again a time effect that seems to play the main role: the "first" arrival of the direct sound (Figs. 4.26 and 4.27) provides the key cues on which our perception system locks in to define the actual tone sensation and tone discrimination (precedence effect).

Visual discriminatory and "lock-in" tasks are performed at a lower level in the "superior colliculus" of the midbrain (Gordon 1972), a nucleus of neurons of the sensory pathway, where incoming signals from the retina and outgoing signals from the cortex can interact with each other, giving feedback instructions from the latter a chance to influence or control the processing of the former. Very likely, a similar mechanism handles the cortex–feedback-controlled lock-in operations in the auditory pathway (Section 2.9).

As happens with two pure tones, there is a minimum difference in fundamental frequency that two complex tones must have in order to be heard out separately. When two complex tones differ in pitch less than the limit for tone discrimination, first order beats (Section 2.4) may arise between all harmonics. If, for instance, both complex tones are out-of tune unisons with fundamental frequencies f_1 and $f_1 + \varepsilon$, respectively, *all* resonance regions will overlap on the basilar membrane and produce beats of different frequencies. The fundamentals will beat with frequency ε, the second harmonics with frequency 2ε, and so on. Only the first few harmonics are important; usually the beats of the fundamental (ε) are the most pronounced. First order beats between corresponding harmonics will also appear if we superpose two complex tones forming other mistuned musical intervals. These are quite different (though equal in frequency) from the second order beats that arise in out-of-tune intervals of *pure* tones (Section 2.6).

<table>
<tr><td>

5.2
The Sensation
of Musical
Consonance and
Dissonance

</td><td>

Consonance and dissonance are subjective feelings associated with two (or more) simultaneously sounding tones, of a nature much less well defined than the psychophysical variables of pitch and loudness, and even timbre. Yet tonal music of *all* cultures seems to indicate that the human auditory system possesses a sense for certain special frequency intervals—the octave, fifth, fourth, etc. It is most

</td></tr>
</table>

[1] Note, however, that language processing is accomplished mainly in the *left* cerebral hemisphere, whereas musical operations involve mainly the right side (Section 5.6).

significant that these intervals are "valued" in nearly the same order as they appear in the harmonic series (see Fig 2.19).

When two complex tones are sounded in unison or in an octave exactly in tune, *all* harmonics of the second tone will pair up exactly with harmonics of the first tone; no "intermediate" frequencies will be introduced by the second tone. This property puts the octave in a very special situation as a musical interval (in addition to the arguments concerning vibration pattern simplicity put forth in the analysis of pure tone superpositions in Section 2.6). The situation changes when we sound a perfect fifth of complex tones (Table 5.1). All odd harmonics of the fifth have frequencies that lie *between* harmonics of the tonic; only its even harmonics coincide. In particular, the third harmonic of the fifth, of frequency $\frac{9}{2}f_1$, lies "dangerously close" to the frequencies of the fourth and the fifth harmonic of the tonic: their resonance regions on the basilar membrane may overlap and either beats or "roughness" may thus appear (Section 2.4), even if the interval of fundamental frequencies is perfectly in tune. By constructing tables similar to Table 5.1, the reader may verify for himself that for such other musical intervals as the fourth, thirds, and sixths the proportion of "colliding" harmonics increases rapidly and moves down in harmonic order. Historically, this effect was thought to be the main cause for the sensations of consonance and dissonance.

Indeed, since the times of von Helmholtz, dissonance was associated with the number, intensity, and frequency of beating harmonics—and consonance with the absence thereof. In other words, it was assumed that for some unspecified reason our auditory system "does not like beats." As a result, it prefers, above all, the perfect unison and the perfect octave because in these intervals all harmonics of the upper tone form matching pairs with harmonics of the tonic. In the fifth, according to Table 5.1, the third harmonic of the upper tone may beat with the fourth and fifth har-

Table 5.1

Tonic	Perfect fifth
f_1	
	$f_1' = \frac{3}{2}f_1$
$2f_1$	
$3f_1$ — — — — —	$f_2' = 3f_1$
$4f_1$	
	$f_3' = \frac{9}{2}f_1$
$5f_1$	
$6f_1$ — — — — —	$f_4' = 6f_1$

monics of the tonic. The increasing proportion of beating pairs of harmonics that appear as one proceeds to the fourth, the sixths and thirds, the sevenths, the seconds, etc., would thus explain the decreasing consonance—or increasing dissonance—of these intervals. This assumption was particularly attractive because—as it is easy to show mathematically—in order to maximize the number of matching harmonics of two complex tones (and hence minimize that of nonmatching ones), it is necessary that their fundamental frequencies f_1 and f_1' be in the ratio of integer numbers, and that these numbers be as small as possible. Indeed, if

$$\frac{f_1'}{f_1} = \frac{n}{m} \qquad n, m: \text{integers} \tag{5.1}$$

then the mth harmonic of f_1' will have the same frequency as the nth harmonic of f_1: $mf_1' = nf_1$ [and so will the $(2 \times m)^{th}$ with the $(2 \times n)^{th}$, etc.] All other harmonics will not match and thus may give beats if their frequencies are near enough to each other. Table 5.2 shows the intervals within one octave that can be formed with small numbers m, n and which are accepted in the Western musical culture as consonances (in decreasing order of "perfection").

On the basis of more recent and more sophisticated monaural and dichotic experiments (Plomp and Levelt 1965) on consonance judgment involving pairs of pure tones and inharmonic complex tones, it became apparent that beats between harmonics may not be the major determining factor in the perception of consonance. Two *pure* tones an octave or less apart were presented to a number of musically naive (untrained) subjects who were supposed to give a qualification as to the "consonance" or "pleasantness" of the superposition. A *continuous* pattern was obtained, that did not reveal preference for any particular musical interval. An example is shown in Fig. 5.3. Whenever the pure tones were less than about a minor third

Table 5.2

Frequency ratio (n/m)		Interval
"Perfect" consonances	1/1	Unison
	2/1	Octave
	3/2	Fifth
	4/3	Fourth
"Imperfect" consonances	5/3	Major sixth
	5/4	Major third
	6/5	Minor third
	8/5	Minor sixth

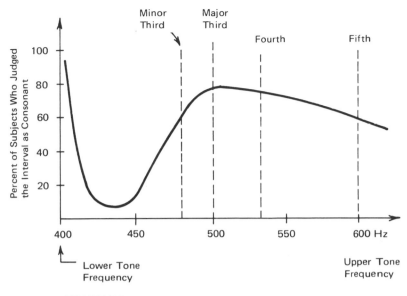

FIGURE 5.3

apart, they were judged "dissonant" (except for the unison); intervals equal or larger than a minor third were judged as more or less consonant, irrespective of the actual frequency ratio.[2] The shape of the curve really depends on the absolute frequency of the fixed tone. All this is related to the "roughness" sensation of out-of-tune unisons and to the critical band discussed in Section 2.4. The results of these experiments may be summarized as follows: (1) When the frequencies of two pure tones fall *outside* the critical band, the corresponding pure tone interval is judged to be "consonant." (2) When they coincide, they are judged as "perfectly" consonant. (3) When their frequencies differ by an amount ranging from about 5% to 50% of the corresponding critical bandwidth, they are judged as "nonconsonant." We shall call an interval of two pure tones under these latter conditions a "basic dissonance."

We now turn back to the musically more significant case of two simultaneously sounding *complex* tones and apply the above results individually to each pair of neighboring upper harmonics. If the total number of pairs that are more or less consonant [see (1) above] and perfectly consonant (2) is weighted against that of basic dissonances (3), a "consonance index" may be obtained for each interval of complex tones (Plomp and Levelt 1965). It can be shown that this index indeed happens to attain peak values for tones

[2] Trained musicians were excluded from this experiment, because they would have been strongly compelled to identify consonances on the basis of training.

whose fundamental frequencies satisfy condition (5.1): the height of the peaks ("degree" of consonance) follows approximately the decreasing order given in Table 5.2. Moreover, in view of the dependence of the critical bandwidth on frequency (Fig. 2.13) *a given musical interval has a degree of consonance that varies along the frequency range.* In particular, moving toward lower frequencies a given musical interval becomes less and less consonant—a fact well known in polyphonic music, where in the bass register mainly octaves and, eventually, fifths are used.

The degree of consonance depends upon the timbre or spectrum of the component tones, i.e., the *relative intensity* of disturbing pairs of upper harmonics. This, too, is well known in music: there are instrument combinations that "blend" better than others in polyphonic music. Even the *order* in which two instruments define a musical interval is relevant. For instance, if a clarinet and a violin sound a major third, with the clarinet playing the lower note, the first dissonant pair of harmonics will be the seventh harmonic of the clarinet with the sixth harmonic of the violin (because the lower even modes of the clarinet are greatly attenuated, Sections 4.4 and 4.5). This interval sounds smooth. If, however, the clarinet is playing the upper tone, the third harmonic of the latter will collide with the fourth harmonic of the violin tone, and the interval will sound "harsh."

Recent studies of the perception of musical intervals invoke periodicity detection (Houtsma and Goldstein 1972), or the mechanism of spatial excitation pattern recognition (Terhardt 1974), as responsible for consonance perception. In the latter theory, it is postulated that tonal music is based essentially upon the pattern recognition mechanisms that operate in the auditory system (Section 4.8). One of these—the central pitch processor—responsible for the extraction of a single pitch sensation from the complicated activity distribution elicited by a musical tone, acquires knowledge of the specific relations that exist between the resonance maxima (and ensuing foci of neural activity evoked by the lower six to eight harmonics of such a tone (Section 4.8). The corresponding primary pitch intervals (octave, fifth, fourth, major third, minor third) thus become "familiar" to the central processor of the auditory system, and convey *tonal meanings* to all external stimuli whose (fundamental) frequencies bear such relationships (Appendix II)[3]. According to this theory, both minimum

---

[3] Quite generally, the hypothesis that the central pitch processor is a neural unit that must *learn* to extract meaningful information from complex input signals through repetitive exposure to natural sounds (Terhardt, 1972; Section 2.9), if demonstrated to be correct, could have

roughness *and* tonal meaning play a determining role in the sensation of subjective consonance. However, in view of the phenomenon of primary pitch shift of individually perceived harmonic components (p. 138), both of these principles may impart conflicting instructions, and, in actual musical situations, force the central processor to a compromise (Terhardt 1974). The preferred "stretched" tuning of pianos (as compared to the equally tempered scale, Section 5.3), and the observed fact that the upper note of a melodic interval of successive tones is preferentially intonated sharp (Section 5.4), may both be a result of this "compromise."

There are more complicated factors that influence the sensation of consonance, most notably experience and training and the ensuing prejudice (i.e., musical tradition) It is interesting to note that, historically, musical intervals as explicit harmonic ingredients have been gradually "accepted" in Western civilization in an order close to the one given in Table 5.2. This seems to point to a gradual refinement and tolerance of our auditory processing ability. Of course, this was not the result of biological evolution but, rather, that of a sophistication of the *learning* experiences to which humans were being exposed as time went on. This development, as that of civilization as a whole, went on stepwise, in "quantum" jumps—it always took the mind of a genius-revolutionary to introduce daring innovations the comprehension of which required new and more complex information-processing operations of the brain, and it was the charisma of the genius that was needed to persuade humanity to learn, and thus to accept *and preserve*, these daring innovations.

So far we have been considering musical intervals smaller than, or equal to, the octave. For large intervals (e.g., C_3-G_5) it is customary to project the upper tone down by octaves (G_5-G_4-G_3) until an interval smaller than one octave is obtained (C_3-G_3). The degree of consonance of that latter interval is then considered "equivalent" to that of the original one. This cyclical property of intervals that repeats within successive octaves has been called the *chroma* of musical tones. It is a basic property that assigns an equivalent rank-

---

far-reaching impact in many ways. In music, for instance, one may set out and try to *relearn* a whole new set of "invariant" characteristics pertaining to, say, a given class of *in*harmonic tones—with the chance of *building entirely new tonal scales and schemes* thereupon (Terhardt 1974). On the more practical side, this intrinsic learning capability would inject additional hope to the present-day efforts of developing electronic prostheses for the deaf, based on *microelectrode implants in the acoustic nerve*. Whereas the spatial activation pattern of these implants is extremely difficult to predetermine, the interpretation of the elicited excitation patterns may well be *learned* by the patient's central processor.

ing to all tones whose pitch differs by one or more octaves, and which makes us call their notes by the same name. What is responsible for this curious cyclic character of musical tones, which repeats every octave (every time the frequency doubles)? It obviously is related to the key property of the almighty octave—that of having all its harmonics coincident with upper harmonics of the fundamental. There is no other musical interval with this property (except the unison, of course). Quite generally, the existence of the chroma, i.e., the fact that pitches differing by an octave have a degree of "similarity" that is considered identical to that of the unison, indicates that the pattern recognition process in our auditory system must respond in some "special," perhaps simplified, way when octaves are presented. Note again that the octave is the first interval in a harmonic series, and that the associated repetition rate is *identical* to that of the lower tone. Any other consonant musical interval (fifth, fourth, etc.) has an associated fundamental repetition rate [relations (2.7)] that is *not* present in the original two-tone stimulus. If we remember how the pitch processor might work (Section 2.9 and Appendix II), we realize that, whenever presented with *two* complex tones whose fundamental frequencies f_1 and f_2 are a musical interval apart, the output from the pitch processor should contain two prominent signals representing the pitch of each tone (corresponding to f_1 and f_2), *plus* other less prominent signals representing the repetition rate (2.7) corresponding to the pair of first harmonics f_1 and f_2 and its multiples (Appendix II). Under normal conditions these additional signals are *discarded* as pitch sensations, a process that requires an additional "filtering" operation. Note, however, that this additional operation is not needed whenever an *octave* is presented, because no such third output signal is present (Figure AII.3, Appendix II)! As a matter of fact, the "tonal meaning" mentioned above may be strongly related to the number, intensity and position of "parasitical" signals in the output from the pitch processor. The more complex the multiplicity of these signals (i.e., the more complex the sound vibration pattern), the "lower" will be the tonal meaning of the original tone superposition.

Finally, when *three or more* tones are sounded together it is customary to analyze the resulting chord into pairs of tones and to consider their individual consonance "values." It is obvious that, as more and more complex tones are combined, a more complicated configuration of resonance regions arises on the basilar membrane. Overlap will increase to the extent that sensor cells covering a large extension of the basilar membrane will respond all at the same time. In light of the various pitch theories (Sections 2.9, 4.8, and Appendix II) we may state that in this case, too, the degree of consonance (or dissonance) may be related to both, the proportion of beating harmonics *and* the number, intensity, and position of "parasitical" signals in the output from the pitch processor. Note that the *major triad* is a three-tone combination whose components, taken two at a time, always yield repetition rates that only differ from the tonic by octaves (i.e., have the same chroma).

There is a limit to multitone intelligibility, though. When the vibration patterns are randomized (i.e., their periodicity is destroyed), or when their complexity exceeds a certain threshold, the neural processing mechanism simply gives up: no definite pitch and timbre sensations can be established. The ensuing sensation is called *noise*. Any non-periodic pressure oscillation leads to a noise sensation. However, noise can be highly organized. Just as a periodic oscillation can be analyzed into a discrete superposition of pure harmonic oscillations of frequencies that are integer multiples of the fundamental frequency (Section 4.3), aperiodic vibrations can be analyzed as a *continuous* superposition of pure vibrations of *all* possible frequencies. Depending on how the intensity is distributed among all possible frequencies, we obtain different *noise spectra*. Noise plays a key role in the formation of consonants in speech. But it also plays a role in music; the importance of noise components in percussion instruments is obvious. The noise burst detected during the first tenths of a second in a piano and harpsichord tone has been shown to be a key element for the recognition process. The effect of noises of electronically controlled spectra on our auditory perception is being studied extensively with tone and noise synthesizers. A vast new territory in auditory sensations (music??) is now being unveiled, (see also Section 5.6).

5.3 Building Musical Scales

For a purely practical purpose, let us define a scale as a *discrete set of pitches arranged in such a way as to yield a maximum possible number of consonant combinations* (or minimum possible number of dissonances) *when two or more notes of the set are sounded together*. With this definition, and keeping in mind Table 5.2, it is possible to generate at once two scales in an almost unequivocal way, de-depending on whether all consonant intervals are to be taken into account, or whether only the "perfect" consonances are to be considered. In the first case, we obtain the *just scale*; in the second, the *Pythagorean scale*.[4]

[4] We envisage here a scale as a set of tones with *mathematically defined frequency relationships*. This is to be distinguished from the various scale *modes*, defined by the particular *order* in which whole tones and semitones succeed each other.

1. THE JUST SCALE

We start with a tone of frequency f_1, which we call *do*.[5] The first most obvious thing to do is to introduce the octave above, which we denote as *do'*. This yields the most consonant interval of all. The next obvious thing is to add the fifth of frequency $\frac{3}{2}f_1$, which we call *sol*. That yields two new consonant intervals, besides the octave, of frequency ratios $\frac{3}{2}$ (do–sol) and $\frac{4}{3}$ (the fourth sol–do'), respectively. For the next step there are two choices, if we want to keep a maximum number of consonant intervals. They are the notes $\frac{5}{4}f_1$ or $\frac{6}{5}f_1$, which we call *mi* and *mi* ♭, respectively. We choose the first one, *mi*, because this will guarantee a number of consonances of higher "degree." Figure 5.4 shows the resulting intervals, all of them consonant. The notes do–mi–sol constitute the *major triad*, the building stone of Western music harmony (our second choice $\frac{6}{5}f_1$ or mi ♭ would have yielded a *minor triad*).

We may continue to "fill in" tones, in each step trying to keep the number of dissonances to a minimum and the number of consonances (Table 5.2) to a maximum. We end up with the *just diatonic scale* of seven notes within the octave (Fig. 5.5). These seven notes can be projected in octaves up and down to form a full diatonic scale over the whole compass of audible pitch. Note in Fig. 5.5 the two intervals of quite similar frequency ratios $\frac{9}{8}$ and $\frac{10}{9}$, representing *whole tones*. The interval $\frac{16}{15}$ defines a *semitone*. With the notes of this scale, taken in pairs, we can form 16 consonant intervals, 10 dissonant intervals (minor and major sevenths, diminished fifth, whole tones, semitones) and —rather unfortunately—*two "out-of-tune" consonances*: the 1.5% too sharp minor third re–fa ($\frac{32}{27}$) and the 1.9% too

[5] The solfeggio notation do–re–mi–fa–sol–la–ti–do is used here to indicate *relative* position in a scale (i.e., the chroma), not actual pitch.

FIGURE 5.4

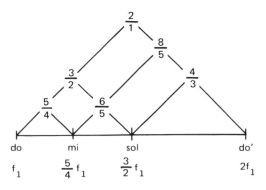

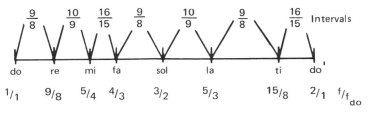

FIGURE 5.5. The just diatonic scale.

flat fifth re–la ($^{40}\!/_{27}$). Finally, and perhaps most important, with the just diatonic scale we can form three just major triads: do–mi–sol, do–fa–la and re–sol–ti; two just minor triads: mi–sol–ti and do–mi–la, and an "out-of-tune" minor triad re–fa–la.

Considering the existence of uneven spacings between neighboring notes, it is possible to still implement this scale by parting the larger gaps (whole tones) into two semitones each. Unfortunately, the resulting new intervals get more and more complex (e.g., several kinds of semitones, more out-of-tune consonances), the choices are not unique, and different frequency values result for the so-called *enharmonic* equivalents do ♯–re ♭, re ♯–mi ♭, etc. Seeking to keep the proportion of possible consonances to a maximum, the following notes are introduced: mi ♭ ($^{6}\!/_{5}f_1$), ti ♭ ($^{9}\!/_{5}f_1$), sol ♯ ($^{25}\!/_{16}f_1$), (or la ♭, $^{8}\!/_{5}f_1$), do ♯ ($^{25}\!/_{24}f_1$) and fa ♯ ($^{45}\!/_{32}f_1$). The result in a *chromatic just scale* of 12 notes within the octave.

2. THE PYTHAGOREAN SCALE

We now restrict ourselves to the so-called perfect consonances, the just fifth and the just forth (and the octave, of course), and build our scale on the basis of these intervals alone. We may proceed in the following way. After introducing *sol* we move a just fifth down from do' to introduce *fa* ($^{2}\!/_{3}\times 2f_1 = ^{4}\!/_{3}f_1$). Then we move a just fourth down from sol to obtain *re* ($^{3}\!/_{4}\times ^{3}\!/_{2}f_1 = ^{9}\!/_{8}f_1$), and a fifth up from the latter to get *la* ($^{3}\!/_{2}\times ^{9}\!/_{8}f_1 = ^{27}\!/_{16}f_1$). Finally we fill the remaining gaps by moving a fourth down from la to obtain *mi* ($^{3}\!/_{4}\times ^{27}\!/_{16}f_1 = ^{81}\!/_{64}f_1$) and up a fifth from there, to *ti* ($^{3}\!/_{2}\times ^{81}\!/_{64}f_1 = ^{243}\!/_{128}f_1$). The result is the so-called Pythagorean scale (Fig. 5.6). Notice that there is only *one* whole tone interval, the *Pythagorean whole tone* of frequency ratio $^{9}\!/_{8}$ (equal to the "short" whole tone of the just scale). The interval $^{265}\!/_{243}$ is the *Pythagorean diatonic semitone*.

We can convert this scale into a chromatic one, by continuing to jump up or down in just fourths and fifths. We thus obtain *fa* ♯ (a fourth below ti), *do* ♯ (a fourth below fa ♯), *sol* ♯ (a fifth above do ♯), *ti* ♭ (a fourth above fa) and *mi* ♭ (a fifth below ti ♭). In this way a new semitone appears (e.g.,

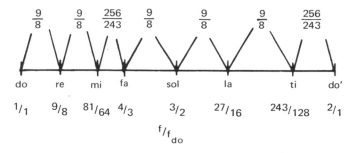

FIGURE 5.6. The Pythagorean diatonic scale.

fa–fa $\sharp$) defined by the odd-looking ratio 2,187/2,048, called the *Pythagorean chromatic semitone*. This whole procedure again leads to enharmonic equivalents of different frequency. In particular, if we continue to move up and down in steps of just fourths and fifths we eventually will come back to our initial note *do*—but not exactly! In other words, we shall arrive at the enharmonic equivalent "ti $\sharp$" whose frequency is *not* equal to that of do' ($= 2f_1$).

So, based on some "logical" principles, we have generated two scales. Each one has its own set of problems. By far the most serious one is the fact, common to both, that only a very limited group of tonalities can be played with these scales without running into trouble with out-of-tune consonances. In other words, *both scales impose very serious transposition and modulation restrictions*. This was recognized as early as in the seventeenth century. There is no doubt, though, that both scales do reveal a quite specific character when music is performed on instruments tuned to either of them. But the type of music that can be played is extremely limited.

3. THE EQUALLY TEMPERED SCALE

It thus became apparent that a new scale was needed, which on the basis of a reasonable compromise, giving up some of the "justness" of musical intervals, would lead to *equally spaced intervals*, regardless of the particular tonality. In other words, a semitone would have the same frequency ratio, whether it was a do–do $\sharp$, a mi–fa or a la–ti $\flat$, and a fifth would be the same whether it was fa–do' or do $\sharp$–sol $\sharp$. This was accomplished in the *tempered scale*, enthusiastically sponsored by none other than J. S. Bach, who composed a collection of Preludes and Fugues ("Das wohltemperierte Clavier") with the specific purpose of taking full advantage of the new frontiers opened up by unrestricted possibilities of tonality change.

In the tempered scale the frequency ratio is the same for

all 12 semitones lying between do and do'. Let us call s this ratio. This means that

$$f_{do\#} = sf_{do}, \quad f_{re} = sf_{do\#} = s^2 f_{do}, \quad \ldots \ldots, \quad f_{do'} = s^{12} f_{do}$$

Since we know that $f_{do'} = 2f_{do}$ (only the octave is kept as a "just" interval!), the twelfth power of s must be equal to 2. Or

$$s = \sqrt[12]{2} = 1.0595 \tag{5.2}$$

This is the frequency ratio for a *tempered semitone*. The frequencies that ensue for the notes of the chromatic tempered scale are *integer powers* of s times f_{do}. Table 5.3 shows the frequency ratios for consonant intervals in all three scales.

It is convenient to introduce a standard subdivision of the basic interval of the tempered scale, in order to be able to express numerically the small differences between intervals pertaining to different scales. This subdivision is used to describe small changes in frequency (vibrato), changes in intonation (pitch), and out-of-tuneness of notes or intervals. The most accepted procedure today is to divide the tempered semitone in *one hundred equal intervals*, or, what is equivalent, to divide the octave in twelve hundred equal parts. Since what defines a musical interval is the *ratio* of the fundamental frequencies of the component tones (not their difference), what must be done here is to divide the semitone frequency ratio s (5.2) into one hundred equal factors c:

$$\underbrace{c \times c \times c \times \ldots \times c}_{100 \text{ times}} = c^{100} = s$$

In view of relation (5.2) the value of c is

$$c = \sqrt[100]{1.0595} = 1.000578 \tag{5.3}$$

The unit of this subdivision is called a *cent*. To find out how many cents are "contained" in a given interval of arbitrary

Table 5.3

Interval	Just scale		Pythagorean scale		Tempered scale	
	Ratio	Cents	Ratio	Cents	Ratio	Cents
Octave	2.000	1200	2.000	1200	2.000	1200
Fifth	1.500	702	1.500	702	1.498	700
Fourth	1.333	498	1.333	498	1.335	500
Major third	1.250	386	1.265	408	1.260	400
Minor third	1.200	316	1.184	294	1.189	300
Major sixth	1.667	884	1.687	906	1.682	900
Minor sixth	1.600	814	1.580	792	1.587	800

frequency ratio r, we must determine how many times we must multiply c with itself to obtain r:

$$c^n = r \qquad (5.4)$$

n is then the "value" of r expressed in cents. By definition, one tempered semitone is 100 cents, a tempered whole tone (s^2) is 200 cents, a tempered fifth (s^7) is 700 cents, etc. To find the value in cents of any other interval we must use logarithms. Taking into account the properties described in Section 3.4, we take logarithms of relation (5.4): $n \log c = \log r$. Hence

$$n = \frac{\log r}{\log c} = 3{,}986 \log r \qquad (5.5)$$

Using this relation, we find the values in cents for the various consonant intervals that have been given in Table 5.3.

5.4 The Standard Scale and the Standard of Pitch

The tempered scale has been in use for more than 200 years and has de facto become the standard scale to which all instruments with fixed-pitch notes are tuned. Since its inception, though, it has come under attack on several occasions—until this day. The main target of these attacks is the "unjustness" of the consonant intervals of the tempered scale, particularly the thirds and sixths (Table 5.3), which do, indeed, sound a little bit out-of-tune when listened to carefully and persistently, especially in the bass register.

Let us critically compare the scales discussed in the previous section with each other. There is no doubt that for *one given tonality* the just scale is the "theoretically" perfect scale, yielding a maximum possibility of combinations of just or "pure" (i.e., beatless) intervals. For this reason, the just scale should indeed be taken as a sort of reference scale; this is precisely why we have introduced it in the first place. But the big question is: Does our auditory system really care for just, beatless intervals? And then: Would we give up tonality transposition and modulation possibilities in favor of obtaining these "pure" intervals? A 200-year history of music has answered these questions unmistakably with a loud and clear *no!* So the just scale is ruled out.

The Pythagorean scale may be one step forward in the right direction (while fifths and fourths are kept as just intervals, thirds and sixths are slightly out-of-tune, Table 5.3), but it still does not allow unlimited transposition and modulation possibilities. There have been other scales, introduced as minor modifications of the Pythagorean scale, that we will not even mention here, however. None has succeeded in preventing the equally tempered scale from being universally accepted.

There have been attempts to settle *experimentally* the question of which scale is really preferred (setting tonality modulation capability arguments aside). There are two possible approaches (1) Use fixed-frequency instruments (piano, organ) and carefully compare the subjective impressions of a given piece of music played successively on two instruments of the same kind, respectively tuned to different scales. The piece of music, of course, should be very simple, without modulations into distant tonalities. And the instrument really should be one with nondecaying tones (such as the organ) in order to bring out beats or roughness more clearly. (2) Measure experimentally the average frequencies of pitch-intonation chosen by singers or by performers of variable-pitch instruments (strings), and determine whether they prefer one scale to another.

The second approach is more appropriate for yielding quantitative results. Recent electronic instrumentation has made possible very precise instantaneous frequency measurements on performers. The musical intervals to be watched closely are the major third and the major sixth, for which the differences between scales are most pronounced (Table 5.3). Notice, in particular, that the upper note in both of these intervals is flat in the just scale and sharp in the Pythagorean scale (with respect to the tempered scale). The experimental results very convincingly show that, on the average, singers and string players perform the upper notes of melodic intervals with *sharp* intonation (Ward 1970). This seems to point to a preference for the Pythagorean scale. However, one should not jump to conclusions: The same experiments revealed that also fifths and fourths and even the almighty octave were played or sung sharp, on the average![6] Rather than revealing a preference for a given scale (the Pythagorean), these experiments point to the existence of a previously unexpected *universal tendency to play or sing sharp the upper notes of all melodic intervals*. This "stretched" intonation could be caused by the primary pitch shift of the harmonic components of a musical tone (Sections 4.3 and 5.2), which leaves a "slightly wrong" record in the central pitch processor. Furthermore, a perhaps even more significant result of these experiments is that individual fluctuations of the pitch of a given note during the course of a performance are very large. This includes vibrato as well as variations of the average pitch of a given note when it reappears throughout the same piece of music. In these pitch fluctuations of a given written note, a frequency range is scanned that goes far beyond the frequency differences between different scales—actually, it

---

[6] A reciprocal effect exists: just melodic intervals are consistently judged to sound flat (Terhardt and Zick 1975).

makes the latter completely irrelevant! Quite generally, all these results point to the fact that *musical intervals are perceived in a categorical mode*, with actual fluctuations being easily ignored by the primary processing mechanism.

So far, we have been dealing with intervals, i.e., frequency ratios. What about the absolute frequencies per se? Once a scale is adopted, it is sufficient to prescribe the frequency of only one note; it makes no difference which one. However, if musical instruments of fixed frequencies are to be easily interchangeable all over the world, this has to be done on the basis of an international agreement. One has prescribed for the "middle A" of the piano (A_4) a fundamental frequency of *440 Hz*. Different "regional" standard frequencies had been in use since the tuning fork became available in the seventeenth century. Over the last two centuries, there has been a gradual rise of the "standard" frequency from about 415 Hz to as high as 461 Hz.[7] We can only hope that the present standard will, indeed, remain constant.

In the tempered scale, all intervals of the same kind (e.g., fifths, major thirds, etc.) are exactly "the same thing," except for the actual pitch of their components. A melody played in C major is in no way different from the same tune played in D major (except for the pitch range covered). Absolute "key colors" or different "moods" of certain tonalities have no psychoacoustical foundation, as experiments have shown (Corso 1957). There can be slight differences in the sounds of various keys, though, due to *physical* circumstances: the greater occurrence of black keys in the piano (which are struck in a slightly different way) for certain tonalities,[8] the greater occurrence of open strings for certain tonalities in string instruments, or the effect of the fixed frequency resonances or range of formants (Section 4.3) in soundboards and other resonance bodies.

One final word is in order on *absolute pitch perception*. The few persons endowed with the ability to recognize or to vocally reproduce a given note in absolute manner (this is also called "perfect" pitch) are usually greatly admired. It happens that *we have been trained to pay attention to, and to store in the long term memory, only relative pitch intervals,* because this is the information most relevant to a

---

[7] This has a serious consequence for famous historical instruments that are still in use today. For instance, a Stradivarius violin, originally built for a standard pitch of, say, $A_4 = 415$ Hz, today has to be tuned higher, which means *higher tension* for the strings [relation (4.3)]. This alters the quality (spectrum) of the tone. A baroque organ, also built for $A_4 = 415$ Hz, when retuned to the higher pitch of $A_4 = 440$ Hz has to have its flue pipes partly cut, in order to shorten their effective length (relation 4.6).

[8] For example, Chopin's piano music!

"musical message." In other words, our brain has been trained to interpret and store a melody as a sequence of pitch *transitions* rather than pitch "values"; the information on absolute pitch, although reaching our brain, is discarded as nonessential in the long–term assimilation process. It can, however, be retained by all normal persons during short intervals of time, ranging from 10 seconds up to a few minutes (Rakowski 1972). It is quite possible that "perfect pitch" could be learned at an early stage of mental education and retained thereafter. These considerations and the surprising excursions in pitch detected electronically in live music, which seem to pass unnoticed by the listener, suggest the need for an operational re-definition of pitch as *the subjective correlate of each one of the acoustical events in a musically meaningful sequence of tones* (Houtsma and Goldstein 1972). It should come to no one's surprise that this is the definition that has been used—implicitly or explicitly—by musicians all the time!

5.5 Why Are There Musical Scales and Why Do We Experience Musical Sensations?

Our ear is sensitive to sound waves over a wide range of frequencies. We can detect very minor changes in frequency; the *jnd* is typically only 0.5% or less (Fig. 2.9). Yet our Western music (and that of most other cultures) is based on scales, i.e., tone transitions and tone superpositions that differ from each other by more than 20 times the limen of our frequency resolution capability. Why don't we make music with continuously changing pitches that sound like, for instance, the "songs" of whales and dolphins (which have a very sophisticated acoustical communications system based on continuous frequency "sweeps")? Why does pitch always have to "jump" in discrete steps?

There are no simple answers to these questions. First, let us remember that a given musical tone has to last a certain minimum period of time in order to be processed by the brain (Section 3.4). This probably has prevented sweeping tones from becoming basic and lasting elements of music. Second, let us note that different musical cultures are using or have used different scales—thus scales are somehow related to, or were influenced by training and tradition. Historically, the existence of scales has been justified on the basis of consonance. This would imply that scales appeared in connection with polyphonic music. However, scales had already been in use when melodies were only sung (or accompanied) monophonically in unison (or, at the most, in octaves or fifths). Quite generally, we believe that there are scales because it is easier for the brain to process, identify, and store in its memory a melody that is made up of a time sequence of discrete pitch values that bear a certain relationship to each other, somehow given by the "familiar" harmonic series, rather than one that sweeps continuously up and down over all possible frequencies, and whose processing, identification, and storage in the memory would require far more information "bits" than a discrete sequence.

The explanation of the existence of scales, i.e., discrete tone sequences, has also been attempted on a *dynamic* basis of tone–tone relationships in time, i.e., based upon *melodic* rather than harmonic intervals. This line of thought is based on the musically so important, but psychophysically almost totally unexplored field of the sensations of "direction" of a two- (or more) tone sequence, of dominance of a given pitch therein and of "return" to that leading pitch (also called "finality"). For instance, we tend to assign a "natural" direction to a two-tone sequence that is upward (in pitch) if the tones are a semitone apart, and downward, if they are a whole tone apart. In both cases, we assign a dominance to the second tone; the "natural" sense is then equivalent to the direction toward, the contranatural to the direction away from, the leading tone. Similarly, a sequence like C-G-C-G-C-G- . . . . . "begs" to be ended on C, whereas the sequence C-F-C-F-C-F- . . . . . "cries" for an F as the terminating note. And if we listen to E-G-E-G- . . . . . neither of the components is satisfactory as an ending—we want to hear C! The whole diatonic orientation of Western music listening is based on these effects, and so is the "meaning" (the tonal "Gestalt") of musical messages (Section 1.3).

At the beginning of this century, Lipps (1905) and Meyer (1900) attempted to "explain" the preference for certain melodic endings and tonic dominance in terms of numerical properties of the frequency ratios of a melodic interval. In the above examples, the dominating tone is that one whose frequency corresponds to a power of two in the integer number ratio. For instance, $f_G/f_C = \frac{3}{2}$; $f_F/f_C = \frac{4}{3}$; $f_C/f_B = \frac{16}{15}$. Later investigations, however, have tended to attribute these effects mainly to cultural conditioning (Lundin 1967). Still, the question remains: Why did these and not any other preferences emerge? It is worthwhile to note, in this connection, that when the musical intervals C-G, C-F, and E-G of the above examples are thought of as neighboring tones of a harmonic series, the fundamental note of that series happens to give the leading pitch as determined by the sense of return (C, F, C, respectively). Again, this may be dictated by the "familiarity" of the harmonic interrelations acquired by our central pitch processor (Sections 2.9, 4.8, 5.2, and Appendix II).

Another perceptual phenomenon related to time sequences of tones, of importance to music, is that of melody *fission*. If a melody is played in which the tones succeed each other fast with melodic intervals of several semitones alternating up and down, coherency is lost and *two* (or more) independent melodic lines are perceived. In this case *our brain tends to group the tones according to their proximity in pitch, rather than their contiguity in time.* This effect has been profusely used, especially during the baroque period, to make it possible to play multipart music on a single-tone instrument. A detailed study and comprehensive review of this and other related time-sequence phenomena has been published recently by van Noorden (1975).

We may venture the prediction that all these effects probably arise in the neural system's "minimum-effort way" of operating: in the identification process of musical (and other sensorial) messages, the system first discards all but a certain minimum of information cues. If the identification has been successful it proceeds ahead with the next message. If not, it goes back to the short term memory and searches for additional cues. This applies not only to a single one-tone input, but also to the musical message as a whole: *the nervous system tries to use whatever information is available from previous experiences* (e.g., long term memory-stored messages) *to facilitate and even anticipate the identification process of new incoming information.* This "prediction" or "extrapolation" capability—quite generally, perhaps the most essential operation of the central nervous system helping to enhance the chance of survival for all higher living beings (Braitenberg 1974)—has been confirmed through electrophysiological measurements (John 1972 and references cited therein). When a certain event in a previously learned succession of stimuli is expected but does not occur, cortical activity (Section 4.9) appears at a latency similar to that usually evoked by the expected event. This capability was also confirmed in speech-processing studies (Christovich 1962). Therefore, whenever this prediction mechanism fails during the perception of music (e.g., because of an "unexpected" passage or a goof in the performance), the extra work required for reidentification gives a particular sensation of "musical tension" (Roederer 1972).

Quite generally, it is possible that the combined effect of complexity of tone message identification (i.e., the total number of required neural operations per unit time) and the associations triggered in other brain centers might be the ultimate "cause" of the musical sensations evoked by a given musical message. For instance, when the tonality of a musical piece is modulated, the auditory nervous system must quickly build up a new "checklist" for the tonal identification operations expected to come. This rapid buildup, an extra load of probably hundreds of thousands of individual operations, would yield the particular musical sensation related to tonality change. Because this rearrangement probably occurs anyway, involuntarily, even if we know the musical piece by heart, the musical sensation appears every time—quite fortunately for music lovers. *Unmusical* individuals, who are unable to experience these sensations, probably are subjects whose musical message identification mechanism has not had a chance to develop its potentialities to full capability. Hence, although they *hear* everything that highly musical subjects hear, their central auditory system can only handle a reduced amount of non-speech related information in a given time. A tonality change passes unnoticed because even for a fixed tonality their auditory processing mechanism is probably saturated with an information flow that exceeds its processing capacity (see also Section 5.6).

In the above discussion we have left out all aspects related to emotionality as an overall "output" of musical perception. It is most perplexing that such a biologically "irrelevant" input as music is at all capable of eliciting an emotional response. The fact that it does often lead to hedonic experiences is an indication that music processing engages the *limbic system.* This is a phylogenetically old part of the brain, which had evolved from a system handling olfactory information, comprising several structures (hippocampus, amygdala, several thalamic nuclei, etc.). In conjunction with the hypothalamus (the part of the brain that integrates the functions of the autonomic or visceral nervous system and regulates the endocrine system), the limbic system polices sensorial input, selectively directs recent memory storage, and mobilizes motor output, with the *specific* function of ensuring a response that is most beneficial for the self-preservation of the organism in a complex, continuously changing environment. It accomplishes this function by *dispensing sensations of reward and punishment* or pleasure and pain, depending upon the current circumstances. Learned behavior, guided by a serial evaluation of responses for their appropriateness to some specific end, is motivated by a principle of "maximum expectancy of reward" (and/or minimum expectancy of punishment) from the limbic system. Emotionality presumably is related to the integrated balance of the reward and punishment signals dispensed during the *actual,* real-time course of behavioral responses. In animals the limbic system is stimulated mainly by environmentally and somatically triggered neural signals. In humans it can also respond to internally evoked images displayed on the cortex during the process of thinking. In other words, dispensation of reward and punishment in man can be triggered by neural information processing that has no relationship whatsoever to the instantaneous state of the environment and the actual response of the organism to it.

The emotional response to music may well be the result of a limbic-system-controlled drive to exercise at an early age in sophisticated acoustical information-processing operations in anticipation of what is needed for speech perception. This training begins with the attention focussed on the simplest (musical!) forms of human verbal utterances. The drive to train persists, and is accordingly rewarded, in later stages of life, even after speech has been acquired.

Emotionality in music may also be the result of a process of integral comparison of the incoming messages with previous musical experiences and the retrieval of information on, i.e., the "replay" of, the circumstances under which they had been experienced. We shall not proceed any further with this purely psychological topic, except to point out that memory storage of acoustical experiences and associations may start at a very early age—as a matter of fact, in the late intrauterine stage (Benenzon 1971), with the fetus perceiving the mother's heartbeat, voice, and gastric sounds, as well as external sounds, filtered through the amniotic fluid.

We also omitted any reference to *rhythm* as a most fundamental component of music. This is a particularly critical omission, because the appearance of rhythm always seems to have been the first step in the evolution of a given musical culture. The propagation through cerebral tissue of a cyclically changing flux of neural signals triggered by rhythmic sound patterns may somehow enter in "resonance" with the natural "clocks" of the brain that control body functions and behavioral response. These clocks probably work on the basis of neural activity traveling in closed-looped circuits or engrams, or in any other neural wiring schemes that have natural periods of cyclic response. Much more research is needed before we can venture to give a clear-cut reply to the question formulated in the title of this section: why are there musical sensations?

**5.6
Specialization
of Speech and
Music Processing
in the Cerebral
Hemispheres**

In the introductory chapter we alluded briefly to the remarkable division of tasks found among the left and right cerebral hemispheres of the human brain (Section 1.5). Time has come now to elaborate further on this phenomenon, mainly in the light of its relevance to music (e.g., see Scheid and Eccles 1974, and references therein).

The body of vertebrates exhibits a bilateral symmetry, especially with respect to the organs concerned with sensory and motor interaction with the environment. This symmetry extends to the brain hemispheres, with the *left cortex* connected to the *right side* of the body, and vice versa. This crossing mainly pertains to the systems capable of sensing directional dimension such as vision and audition (e.g., see flow chart in Fig. 2.25), and to the efferent motor control of legs and arms. It probably developed because of the need to keep together within one cortical hemisphere the interaction mechanisms connecting incoming information and outgoing motor instructions regarding events from the same spatial half-field of the environment. The optical image is *physically* inverted in the eye lenses, projecting the right panoramic field onto the left half of the retina and vice versa, in each eye. The *left* halves of both retinas are connected to the *left* visual cortex, in order to reunite in one cerebral hemisphere the full information pertaining to the same spatial half-field.

As mentioned in Section 2.9, both hemispheres are connected with each other by the 200 million fibers of the corpus callosum, which thus restores the global unity of environmental representation in the brain. In the afferent auditory pathways there are lower-level connections between the channels from both sides (e.g., Fig. 2.25), through which the left and right side signals can interact to provide information on sound direction.

In the evolution of the human brain, the immense requirements of information processing that came with the development of verbal communication crystallized in the emergence of *hemispheric specialization*. In this division of tasks, *the analytic and sequential functions of language became the target of the "dominant" hemisphere* (on the left side in about 97% of the subjects—Penfield and Roberts 1959). *The minor hemisphere emerged as being more adapted for the perception of synthetic, holistic relations.*[9] That the speech centers are located in one hemisphere (most frequently the left) has been known for over 100 years, mainly as the result of autopsy studies conducted on deceased patients with speech and language defects (comprehension, reading, writing) acquired after cerebral hemorrage (strokes) in one given hemisphere (e.g., Geschwind 1972). More recently, some convincing examples have been documented with studies on "split-brain" patients whose corpus callosum has been surgically transected for therapeutic reasons (e.g., Gazzaniga 1970). For instance, these patients cannot verbally describe any object, written word, or event localized in their left visual field, because the pertinent information, originally displayed on the right-side visual cortex, cannot be transferred to the speech centers by the severed corpus callosum! On the other hand, tests on patients with lesions of the *minor* (right-side) hemisphere have revealed that visual pattern recognition is impaired (Kimura 1963). Similarly, timbre and tonal memory tests (Milner 1967) show impairment of auditory pattern recogni-

[9] It might be suspected that speech processing is located in that hemisphere that also controls the predominantly used hand (e.g., left hemisphere for the right hand). But there seems to be no direct correlation between the predominant right-handedness and left-side speech processing: left-handedness is much more frequent than the 3% of right side speech processing.

tion (Section 4.8). Quite generally, all *non*verbal auditory tasks are impaired in these patients. This supports the indication that *the central mechanisms relevant to the perception of music are located fundamentally in the minor hemisphere* (the temporal lobe thereof). This has been confirmed more recently through experiments involving normal, healthy subjects, with the technique of dichotic listening tests. This technique is based on the fact, mentioned in Section 2.9, that auditory information flowing through the main contralateral auditory channels (Fig. 2.25) *overrides* any conflicting information traveling the ipsilateral route. Hence, when both ears receive conflicting information, the *left* auditory cortex will pay more attention to the *right* ear input (although it also receives information from the left—Fig. 2.25), and vice versa. Indeed, a right-ear advantage was found in speech recognition tasks, and a left-ear advantage for melody tests (e.g., Kimura 1973).

Why are musical messages mainly processed in the minor hemisphere? Is it by "default," because the other hemisphere has specialized in the processing of speech?[10] In that case, of course, the question remains: Why is speech handled by only *one* hemisphere? One may shed some light on this problem, by realizing that the specialization of the cerebral hemispheres is of a much more basic nature, involving *two quite different operational modes*. One mode involves *sequential analysis* of subparts (subparts timewise) of information such as required in language processing. The other involves *spatial integration* or synthesis of momentaneous patterns of neural activity, to accomplish the determination of holistic qualities of input stimuli (e.g., Papçun et al. 1974). However, both modes must coexist and cooperate in order to process information on, and program the organism's response to, the complex human[11] environment. In particular, sequential tasks (like visual scanning) may be necessary for pattern recognition and image construction and, conversely, holistic imaging may be required as a subroutine operation collateral to sequential programming.

Holistic pattern recognition is a most fundamental requirement for animal survival in an environment in which many correlated events occur at the same time in different spatial locations. In contrast, sequential analysis became the fundamental mode for processing language and for controlling speech and thought processes—wherein the "events"

---

[10] We should point out that this specialization is not absolute. It has been shown (for instance, with split-brain patients) that in adult persons the minor hemisphere seems to cooperate in normal speech handling the "musical" contents of speech (vowels, tone of voice, inflections).

[11] By "human environment" we mean an environment containing other humans with whom to intercommunicate.

follow one another serially in time. Indeed, communicating and thinking (which is also a form of communication—between one's own brain centers) are brain operations that are fully and fundamentally based on short-term time sequencing. Animal communication capabilities are infinitesimal compared to man's verbal capabilities, animals do not think like man does[12]—and animals do not show any hemispheric specialization.

Because music is preferentially handled by the minor hemisphere, does this mean that music mainly involves synthetic operations of holistic quality recognition? Regarding complex tone recognition, this indeed seems to be in agreement with the recent theories of pitch perception (Sections 4.8 and 4.9). *The holistic quantity in a musical stimulus is the momentaneous distribution of neural activity* (corresponding to the resonance maxima on the basilar membrane), leading to complex tone pitch (Section 4.8), to multiple-tone discrimination (Section 5.1), to consonance (Section 5.2), and tonal return (Section 5.5). Another quantity is the relative distribution of the *amount* of activity, given by the power spectrum, leading to timbre and tone source identification (Section 4.9). We may identify here a good formal analogy with vision: the incoming sound pattern (in time) is "projected" as a pattern in space on the basilar membrane—the result is a *spatial* image, much like the spatial image projected on the retina. From there on, both systems operate on their respective inputs in formal analogy, eventually leading to musical and to pictorial sensations.

An apparent paradox emerges when we consider *melodies* and the time dependence of musical messages. Wouldn't they require sequencing, i.e., *dominant* hemisphere operations? This is not necessarily so. Speech (and thinking) involves *short-term* sequencing that mainly engages the short term memory in all auxiliary subroutine operations. Melodies and musical messages on the other hand mostly exceed the storage time and the capacity of the short term memory and associated information transmission channels. This leads to the generally accepted idea that *prima facie our brain recognizes the typical musical messages as being of holistic nature*, long term patterns in time, rather than short term sequences. The phenomenon of melodic fission (p. 162) is a most convincing example of this. Expressed in other words, music seems to be recognized by our brain as the representation of *integral, holistic auditory images* (the harmonic

---

[12] Here we define "thinking" as an operation in which information is retrieved from memory *without* external stimulation, intercompared, rearranged, reanalyzed or recorrelated, and re-stored as a *modified or corrected version* of the original representation.

structure), *whose* (long term) *succession in time bears in itself a holistic Gestalt value* (the melodic contours).

As discussed above, the left temporal lobe is specialized in processing "verbal" inputs, and the right in handling "musical" inputs. An interesting result is the following: even nonsense speech, played backward on a tape recorder, is preferentially handled by the *dominant* hemisphere (Scheid and Eccles 1974). This only confirms the short term character of the sequencing operations relevant to speech, mentioned above: they are concerned with a stage in the processing of acoustical information that comes *prior* to the recognition of conceptual content. On a similar line, recent dichotic listening tests (Bever and Chiarello 1974) have shown that when *musically experienced* subjects focus on short term fine structures of tonal successions in a melody, the corresponding analytical processing strategies are mainly handled by the *dominant* hemisphere. A reverse situation arises with a *sung text*. It has been found that patients with severe speech impediments (aphasias) are able to sing with clearly understandable words a song learned before the trauma had arisen—but they are unable to *speak* those same words. This suggests that speech serving musical expression is processed preferentially in the *minor* hemisphere. Language in these situations (and probably also in poetry) has much of a holistic and synthetic quality (Scheid and Eccles 1974).

All this is quite germane to the understanding of the evolution of Western music. In a broad sense, we may depict this evolution as a gradual transition between two extreme configurations. At one extreme we find highly structured, clearly defined, emphatically repeated, spatial (harmonic) and temporal (melodic) sound patterns, each one of which bears a value as an unanalyzed whole (e.g., a given chord and a given voice or chord progression, respectively). At the other extreme (to which we are now heading), we identify tonal forms whose fundamental value is recognized in the momentaneous state of the short term temporal sound signatures. In the light of what we have said above about hemispheric specialization, we may speculate that these two extreme configurations are intimately related to the two distinct processing strategies of the human brain. Only the future will tell whether the current trends in music merely represent a more or less random effort to "just break away" from traditional forms (which in part had emerged quite naturally as the result of physical properties of the human auditory system), or whether these trends can be channeled into a premeditated exploration and exploitation of vast, still untested, processing capabilities of the central nervous system.

Finally, we should come back to the question of general

symmetry of vertebrate bodies and brains. In humans, a remarkable *asymmetry* has been identified,[13] showing that the *planum temporale*—a cortical area that plays a key role in auditory processing—is significantly larger in the left temporal lobe than in the right, in 65% of all examined cases (Geschwind and Levitzky 1968). In 24% of the cases there was no significant asymmetry, and in the remaining 11% the asymmetry was reversed. No such asymmetry was found in the brains of nonhuman primates. It is highly suggestive that this asymmetry bears some relationship wtih the asymmetry of the tasks carried out by both hemispheres. Although the figure of 65% is much lower than the percentage of speech representation in the left hemisphere (about 97%), it is quite possible that the relatively large right–side planum temporale in the remaining 32% is indication of an inborn greater capacity of nonverbal sound processing. Moreover, it has been suggested (Scheid and Eccles 1974) that *the enlargement of the right planum temporale could be a measure of inborn musical ability.* This hypothesis, if confirmed statistically through systematic postmortem examinations,[14] would give an anatomical foundation to the hereditary transmission of musicality.

[13] Reported to exist even in babies and fetuses (Scheid and Eccles, 1974).

[14] Musicians: donate your remains to a worthy cause!

Appendix I:
Some Quantitative Aspects
of the Bowing Mechanism

Let us consider an idealized situation: a very, very long string, bowed at a point A with an infinitesimally thin bow (Fig. AI.1). The bow moves with a speed b in the upward direction. Let us furthermore assume that right from the beginning (time t_0) the string sticks to the bow. This means that the point of contact A (we really should say the segment of contact) also moves upward with the same speed b. The result will be a deformation of the string in the form of a transverse wave that will propagate away from point A as shown in Fig. AI.1 for various instants of time t_1, t_2, t_3. Since transverse waves propagate with a speed V, given by relation (3.3), which happens to be much higher than any reasonable bowing speed b, the slope b/V of the kinked portions of the string AP, AQ will in reality be extremely small. Under these conditions, the transverse force F applied by the bow (not

FIGURE AI.1

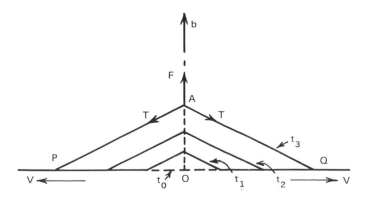

to be confused with the bowing pressure which would be directed *perpendicular* to the paper) holds the balance with the projections along *OA* of both tension forces *T*. This means that $F = 2T\, b/V$. In order to actually have a regime of static friction, with the string sticking to the bow, the force *F* must be less than a certain threshold F_s, called the limit of static friction. Experimental results show that this limit is proportional to the bowing "pressure": $F_s = \mu_s P$. μ_s (Greek letter mu) is the *coefficient of static friction*; it depends on the "roughness" of the contact surfaces (in this case, on the amount of rosin on the bow's hairs). The condition for "sticking" is then $F = 2T\, b/V < \mu_s P$.[1] Likewise, the condition for slipping will be $F = 2T\, b/V > \mu_s P$. Since the quantities *V*, *T*, and μ_s are constant parameters for a given string, we may summarize both expressions in the following, more physical way:

Quantity controllable by the player	Quantity fixed for each string	Type of string motion with respect to the bow	
$\dfrac{b}{P}$	$\begin{cases} < \dfrac{\mu_s V}{2T} \\[2ex] > \dfrac{\mu_s V}{2T} \end{cases}$	sticking slipping	(AI.1)

Notice in relations (AI.1) that what matters is the *ratio* of bow velocity to bow pressure, not *b* or *P* alone. The ratio *b/P* thus defines the nature of the motion of the bowed string.

What now, if the string is slipping from the very beginning? [bottom relation in (AI.1)] In that case the speed *v* of the bowing point *A* of the string will be less than (or may even be oppositely directed to) the speed of the bow, *b*. We have a regime of *dynamic* friction, in which it turns out that the force $F = 2T\, b/V$ (Fig. AI.1), while still proportional to *P*, now also depends on the *relative speed* $b - v$ between bow and string (speed of slipping). We write this in the form $F = \mu_D P$, where μ_D is the coefficient of *dynamic friction*, which now depends on the relative velocity $b - v$ (is a *function of* $b - v$). Thus, during the slipping regime:

$$\frac{b}{P} = \mu_D \frac{V}{2T} \tag{AI.2}$$

If we knew the dependence of μ_D with the speed of slipping $b - v$, we could use expression (AI.2) to determine the speed *v* of the contact point *A* of the string. Again, this relation is governed by the ratio *b/P*. But notice carefully: what is determined by this ratio is the difference $b - v$, i.e., the

---

[1] $<$ means "less than," $>$ means "greater than."

speed of the string *as seen by the bow*. The larger b is, the larger will be v, for a given value of b/P. While this ratio determines the *nature* of the string motion [sticking versus slipping, relations (AI.1)], the bow speed determines the actual speed of the string (for a given b/P). Thus if one increases bow speed, but *at the same time* increases bow pressure so to hold their ratio constant, the nature of the string motion will not change at all—only its velocity will increase linearly with b. This will lead to an increase of the amplitude, i.e., intensity of sound, in the real case. In other words: *the amplitude of the vibration of a bowed string (loudness of the tone) is solely controlled by the bow velocity, but in order to maintain constant the nature or type of the string motion (timbre of the tone), one must keep the bowing pressure proportional to the bowing speed.*

Although relations (AI.1 and AI.2) had been derived from an "infinitely" long string, these results are still true for the "real" case of a string of finite length. Let us now consider a little more realistic case: a string of finite length L, bowed with an infinitesimally thin bow at the midpoint O (Fig. AI.2). In this figure we show schematically the shape of the string as we start bowing (again, the slopes are *highly* exaggerated). v is the speed of the midpoint [we may either have slipping $(v < b)$, or sticking $(v = b)$]. Notice that at time $t_4 = L/2V$ the first "wave" (of slope v/V) has reached the string's end points. There, the wave is reflected and superposed with the ongoing initial wave, unfolding the "kinked" shape shown for the times t_5 to t_7. Then, at $t_8 = L/V$ something new hap-

FIGURE AI.2

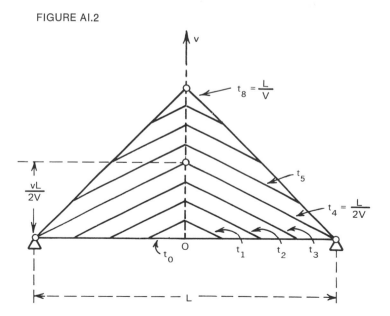

pens (Fig. AI.2): *the slope suddenly changes at the bowing point*. This changes the expression for the force F and a new regime may arise (e.g., slipping, if previously there was sticking). We cannot continue this discussion without running into considerable mathematical complexities (Keller 1953). Just notice that these crucial changes in shape (whenever the wave is reflected at the fixed end points) always occur at times that are *integer multiples* of L/V, a quantity that is completely independent of the bowing mechanism. Actually, the inverse of L/V appears in the expression (4.3) of the fundamental frequency of the vibrating string. The reader may thus envisage how that frequency (and all upper harmonics) may indeed be excited (and maintained) by the bowing mechanism. He may also infer from Fig. AI.2 (with a little extra imagination) that in its real vibratory motion, a *bowed string always has an instantaneous shape that is made up of sections of straight lines;* this result has been verified experimentally long ago.

Appendix II:
Some Quantitative Aspects
of Recent Central Pitch
Processor Models

In this Appendix we follow up on our discussion of the perception of the pitch of complex tones given in Sections 2.9 and 4.8. In particular we show how, with just a little algebra, the model of "template fitting" (Goldstein 1973) can explain some quantitative characteristics of complex-tone pitch perception (the other models described in the abovementioned sections lead to quite similar quantitative results, but are not so easily accessible to a simple mathematical treatment). In the second part we shall speculate about neural models that may accomplish the functions of a central pitch processor.

Goldstein's model (Section 2.9) is based on the assumption that the neural information on the spatial positions of resonance maxima on the basilar membrane is not sharply defined, fluctuating statistically around average values. The match of the "template" that ultimately will lead to the subjective pitch sensation is one which minimizes the differences between the built-in template "values" and those of the real signal. To illustrate how such a maximum likelihood estimation would work, let us consider a two-tone stimulus of frequencies $f_a = 1,000$ Hz and $f_b = 1,200$ Hz. These are the *exact* fifth and sixth harmonic frequencies of a fundamental $f_1 = 200$ Hz. Our "template" is represented by a harmonic set of frequencies $f_1, 2f_1, \ldots nf_1 \ldots$ whose fundamental f_1 can be changed arbitrarily. The matching process consists of finding a frequency f_1 for which two *successive* harmonics nf_1 $(n+1) f_1$ coincide with, or are as close as possible to, the input tone frequencies f_a and f_b. At this stage it is irrelevant what the order of these two harmonics is (what the value fo n is), provided that the match is the best of all possible ones (the differences $[nf_1 - f_a]$ and $[(n+1)f_1 - f_b]$ are both as small as possible). In our example only one match is best: for $n=5$ and $f_1=200$ Hz, both frequency differences are exactly zero—the match is perfect. The reader can easily verify that there is *no* other fundamental frequency f_1, nor any other value of n, that can give a perfect match. Note carefully that a given match requires determination or estimation of *two* values: harmonic order n and fundamental frequency f_1. The thesis of the theory is that this latter frequency would then correspond to the actual, single, subjective pitch sensation evoked by the two-tone stimulus. All this should work in a quite similar way when more than two signals belonging to a perfectly harmonic series (as in a complex tone) are present, and should yield the frequency f_1 corresponding to the pitch of the complex tone (irrespective of whether the fundamental f_1 is actually present in the original stimulus).

Let us now shift the input frequencies to the *inharmonic* pair $f_a=1{,}050$ Hz and $f_b=1{,}250$ Hz. There is *no* harmonic series of which these two are neighboring components. How does template matching work in this case? According to the above theory (an oversimplified version thereof), we must find a pair of values n, f_1 such that the neighboring harmonics nf_1, $(n+1)f_1$ *minimize the root-mean-square relative error*, or, what is the same, maximize the reciprocal quantity $Q(n)$, which we call *quality of fit*:

$$Q(n) = \cfrac{1}{\sqrt{\left\{\dfrac{(nf_1-f_a)^2}{(nf_1)}\right\} + \left\{\dfrac{[(n+1)f_1-f_b]}{[(n+1)f_1]}\right\}^2}} \tag{AII.1}$$

For each value n, there will be one frequency f_1 that maximizes the value of $Q(n)$, which can be easily found through elementary calculus:[1]

$$f_1 = \cfrac{\left(\dfrac{f_a}{n}\right)^2 + \left[\dfrac{f_b}{(n+1)}\right]^2}{\dfrac{f_a}{n} + \dfrac{f_b}{(n+1)}} \tag{AII.2a}$$

A linear approximation of this relation is

$$f_1 \simeq \frac{1}{2}\left[\frac{f_a}{n} + \frac{f_b}{(n+1)}\right] \tag{AII.2b}$$

Inserting this value of f_1 into expression (AII.1) we obtain the actual value of the quality of the fit $Q(n)$:

$$Q(n) = \frac{1}{\sqrt{2}} \left| \frac{1 + \left[\dfrac{n}{(n+1)}\right]\dfrac{f_b}{f_a}}{1 - \left[\dfrac{n}{(n+1)}\right]\dfrac{f_b}{f_a}} \right| \tag{AII.3}$$

---

[1] The reader may indeed verify that this gives the right answer for the above harmonic example, with $n=5$ and $f_1=200$ Hz, for which $Q=\infty$ (perfect fit).

Table AII. 1   Values of f_1 *(AII.2) and* $Q(n)$ (AII.1) (shown in boldface), for a Two-Tone Complex of Frequencies f_a, f_b.

f_b/f_a (Hz)	$n=4$	$n=5$	$n=6$	$n=7$
1,200/1,000	f_1 = 245.1 Hz	200.0	169.1	146.5
	$Q(n)$ = **35**	**∞**	**50**	**29**
1,250/1,050	256.5	209.2	176.8	153.2
	29	**178**	**70**	**35**
1,300/1,100	267.7	218.4	184.5	159.9
	25	**93**	**110**	**42**
1,350/1,150	279.0	227.5	192.3	166.6
	23	**64**	**228**	**53**
1,400/1,200	290.3	236.7	200.0	173.2
	21	**50**	**∞**	**69**

For different values of n, we will obtain different fundamental frequencies f_1, and different fit quality values $Q(n)$. We are thus led to a whole series $Q(n)$, $Q(n+1)$ . . . If among them, *one* stands out as the largest, the corresponding f_1 and n give the best match of all—and f_1 represents the pitch that is heard! In our two-tone example of $f_a = 1{,}050$ Hz, $f_b = 1{,}250$ Hz, we obtain, for $n=4$, $f_1 = 256.4$ and $Q(4) = 29$; for $n=5$, $f_1 = 209.2$ and $Q(5) = 178$; for $n=6$, $f_1 = 176.8$ and $Q(6) = 70$. Clearly, the choice $n=5$ leads to the highest Q value. The corresponding frequency[2] (209.2 Hz) is indeed the subjective pitch that is identified most easily when this inharmonic two-tone complex is presented (see Smoorenburg 1970). Note that $n=6$ and, to a lesser extent, $n=4$ also give a non-negligible quality of fit; this explains the observed fact that the corresponding fundamental frequencies $f_1 = 176.8$ Hz and 256.4 Hz can be identified as "secondary" pitch sensations, although with a little more difficulty. This model thus explains quantitatively the ambiguous or multiple pitch sensations that are elicited by inharmonic tones. If we shift the two-tone stimulus frequencies even further away from harmonicity (but always keeping the same frequency difference of 200 Hz) we obtain the results shown in Table AII.1.

[2] It is important to point out that this frequency is *not* the repetition rate of the inharmonic two-tone complex. That rate still remains at $f_b - f_a = 200$ Hz. Smoorenburg's experiments thus demonstrate that *repetition rate per se is not encoded as the fundamental pitch signal in the case of inharmonic tones.* We must also point out here that *combination tones* of the type (2.5) and (2.6) (Section 2.5) play a role and must be taken into account as *additional* perturbations, before the pitch-matching procedure is applied.

Note that for the pair 1,300/1,100, *two* template matchings are prominent for harmonic orders $n = 6$ and $n = 5$. Note also in Table (AII.1) how the harmonic order n for which the highest value of Q is obtained shifts from 5 to 6, as the center frequency of the pair f_a, f_b is shifted up, while the corresponding best-fit pitch jumps down, from a value lying above 200 Hz to one below 200 Hz. Quite generally, as both f_a and f_b are swept continuously upward in frequency (keeping the difference f_b–f_a constant), the main subjective pitch sensation "oscillates" about the repetition rate given by f_b–f_a (200 Hz), coinciding with the latter in the harmonic positions. Ambiguous or multiple pitches appear most distinctly whenever the two-tone stimulus lies approximately halfway between harmonic situations. In this context one assumes that *the quality of fit Q is related to the "clarity" or intelligibility of the corresponding pitch sensation.*

This fitting method can be extended to multitone stimuli. It is quite illustrative to use such an extended version to find out the predicted pitch or pitches of "weird" combinations, as well as those of, say, superposed complex tones, as in musical chords[3] (see further below). It is interesting to note that this model of pitch extraction via template matching works in formal analogy to the mechanism of pitch adjustment in a wind instrument (p. 124).

A consequence of musical importance is the following. Relations (AII.2) or (AII.3) can be used to show that when a given upper harmonic in a complex vibration gets out of tune ($f_n = nf_1 + \varepsilon$), the effect on the subjective pitch is *very* small (of order $\varepsilon/2n$ or less). Thus, the minor inharmonicities of upper harmonics arising in vibrating strings (p. 97) can only have a second order effect on the resulting subjective pitch.

Let us now speculate on a neural model for the central pitch processor, as a follow-up to the "learning matrix" invoked by Terhardt (1974). This is done here merely as an *academic exercise* and is not intended to represent yet another central pitch theory.

Neural network models are often based on currently known possibilities, but not necessarily on anatomic realities, of the nervous system. Yet modeling is becoming quite an essential tool for the quantitative understanding of neural processing—much in the same way as theoretical physics, which also frequently must deal with oversimplified "armchair" modeling, has long ago been incorporated as an "equal-rights partner" of experimental physics. After all,

---

[3] Great care has to be taken in extending relations (AII.1) and (AII.2) to a multitone case—straightforward summation in enumerator and denominator does *not* yield the correct result, unless the input components are all *nearly* harmonic.

quantitative predictions (Section 1.4) can only be formulated on the basis of quantitative *models*, to the same extent as all quantitative representations of the physical world are model approximations, be they a physical theory, or the representation of the environment in our own brain!

We have sketched in Fig. AII.1 a neural wiring scheme capable of performing the operations needed for pitch extraction and fundamental tracking, based on Terhardt's learning matrix (1974). The horizontal fibers are assumed to conduct the combined neural signals from both cochleas into the *primary* (spectral) pitch processor. This output carries information on each harmonic component of a complex tone, but is assumed to be normally ignored at the higher stages of musical tone processing. These horizontal axons are intercepted by a vertical array of neuronal dendrites (Section 2.8) as shown in Fig. AII.1. We assume that, initially (at birth), the active synaptic connections are distributed as shown for neurons K, L, and M. We further assume that, in order to reach threshold and fire, *each vertical neuron must be activated at many synaptic contacts at nearly the same time*. It is clear from this figure that in an acoustically virgin brain, the output activity distribution from the vertical

FIGURE AII.1

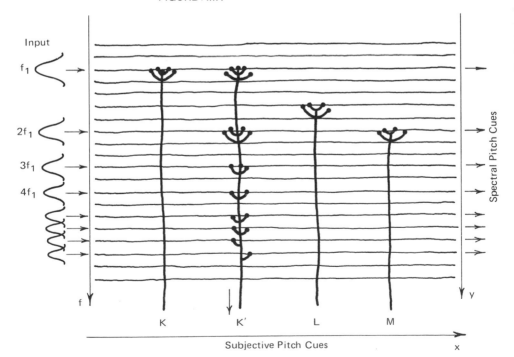

neurons (along the x dimension) would be nearly identical to that of the horizontal fibers (along the y dimension).

Our next assumption, in line with Terhardt's theory, is that as the ears are exposed repeatedly to harmonic tones, *synaptic contacts will also be activated between a vertical neuron and between all those horizontal axons that are most likely to be firing at the same time* (the "essence" of the learning process in the nervous system). When a complex harmonic tone is presented whose fundamental is, say, f_1, neuron K (Fig. AII.1) will initially respond *only* to that fundamental. But as this complex tone stimulus is repeated, the dendrite K will develop active synaptic contacts with all those horizontal fibers whose best frequencies correspond to the higher harmonics of that tone (p. 48). As a result, the vertical neuron emerges "tuned" to *the whole harmonic series* of f_1, as shown for neuron K' in Fig. AII.1. Vertical neurons thus would physically represent a first approximation to the "templates" postulated in Goldstein's theory (see above). Template response will be highest wherever a local best match (local maximum of vertical neuron excitation) is achieved. We finally assume that the location of maxima of output activity from the vertical neurons (along the x dimension, Fig. AII.1) leads to the sensation of subjective or periodicity pitch. After the learning process, this output is indeed quite different from that of the horizontal fibers (in the y dimension). For instance, if enough synapses are activated by the upper harmonics of f_1, *neuron K' will respond, even if the input at the fundamental f_1 is missing* in the original tone. This represents the mechanism of fundamental tracking. The higher the order of the harmonics, the less sharply defined is the "horizontal" input, because of the proximity of the respective excitation maxima (Fig. AII.1). Vertical neurons may thus be led to respond to the "wrong" input signal (one that does not correspond to the fundamental frequency to which its apical dendritic tree has been originally wired). Multiple pitches are hence possible, as we have shown quantitatively in the first part of Appendix AII.

Our model needs some improvements, though. As shown in Fig. AII.1, the "tuned" neuron K' would also respond to all those complex tones whose fundamental frequencies are integer multiples of f_1. It may even fire when only *one* upper harmonic is present. To prevent this undesirable effect from happening, we may introduce an intermediate set of vertical neurons, capable of *detecting coincidences between neighboring harmonics*. This is shown schematically in Fig. AII.2. Now it is the "short" vertical interneurons that are assumed to be the primary learning elements, requiring the simultaneous activation of horizontal fibers corresponding to neighboring harmonics, in order to fire a signal into the

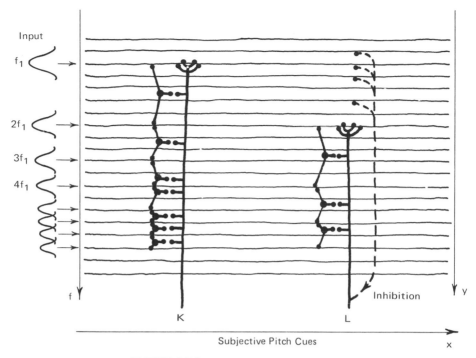

FIGURE AII.2

"long" vertical dendrites. The harmonics of a tone of fundamental nf_1 would not meet the requirement of simultaneous excitation of these interneurons, as the reader can easily verify on the basis of Fig. AII.2.[4]

Unfortunately, one more improvement would be desirable. Even with the coincidence interneurons of Fig. AII.2, a vertical neuron would still fire "unwanted" pulses, when a complex tone of *subharmonic* fundamental frequency f_1/n is presented. Certain upper harmonics of such a tone (of frequency $(m/n) f_1$, with the ratio m/n a whole number) will indeed coincide with harmonics of f_1, to which our neuron is tuned (in Fig. AII.2, the neuron L, tuned to $2f_1$, would fire whenever neuron K is activated). To prevent this from happening we could introduce *inhibitory neurons* in such a

---

[4] The reader should, however, also realize that this scheme of inter-neurons could never be "perfect"—if its synaptic topology emerges from a learning process, there would always be *some* interneurons for neuron K that respond to a signal belonging to a tone of fundamental nf_1. Also, recent experiments (R. Plomp, private communication 1975) show that simultaneity of *neighboring* harmonics does not seem to be an absolute requisite for fundamental tracking (e.g., we hear the correct subjective pitch of tones made up of only odd or only even upper harmonics).

way as to decrease the firing threshold of a vertical neuron whenever frequencies below f_1 are present (in Fig. AII.2, whenever axons are firing above the apical end of a vertical neuron). Or we simply may invoke the increasingly degraded definition of primary excitation peaks, corresponding to higher harmonic order, as well as their upward shift (interval stretching, p. 159), which in the case of *sub*-harmonic input tones of fundamental f_1/n could well account for a greatly reduced excitation probability of a vertical neuron tuned to f_1.

It is interesting to note that at this stage of our model, the response distribution in the x dimension of the vertical neurons (Fig. AII.2) is pretty much what one would expect from a "spatial" autocorrelator as postulated in Wightman's theory (1973). In fact, it is quite straightforward to write a simple program[5] for a digital computer which simulates the operation of the neural model shown in Fig. AII.2.

Some quantitative results are shown in Fig. AII.3 (for simple, but realistic, assumptions on the distribution of primary excitation around each harmonic and on the degradation of response at increasing harmonic order). At the top of each panel, the primary input power spectrum is given (in *linear* scales), corresponding to the superposition of two complex tones forming an octave, fifth, and minor third, respectively. The lower graphs represent the computed neural activity distribution along the x dimension (cf. Fig. AII.2). Note the pronounced peaks corresponding to the fundamental frequencies of each original complex tone. We assume that these peaks are recognized at a higher stage of neural processing and lead to the two principal pitch sensations corresponding to the two-tone complex. The position and, to a large extent, the shape of these primary peaks is independent of the actual power spectra of the component tones, depending only on their fundamental frequencies f_a and f_b. Note also that, following the order of decreasing consonance (Section 5.2), "parasitical" peaks appear at the positions of the repetition rate r and its multiples. These parasitical peaks (which must be deliberately inhibited at some higher stage) are *absent* in the octave. In addition, there is a background activity or "noise level" below the lower pitch tone that increases with decreasing order of consonance.

---

[5] This program makes no assumptions on Fourier transforms or autocorrelation functions. The program simply runs a "template" of harmonic frequencies through a given fundamental frequency domain and counts, for each position, the number of *simultaneous* "excitations" at neighboring harmonic positions. The total number of pairs of simultaneous excitations represents the output (the intensity or probability of activation of the corresponding vertical neuron).

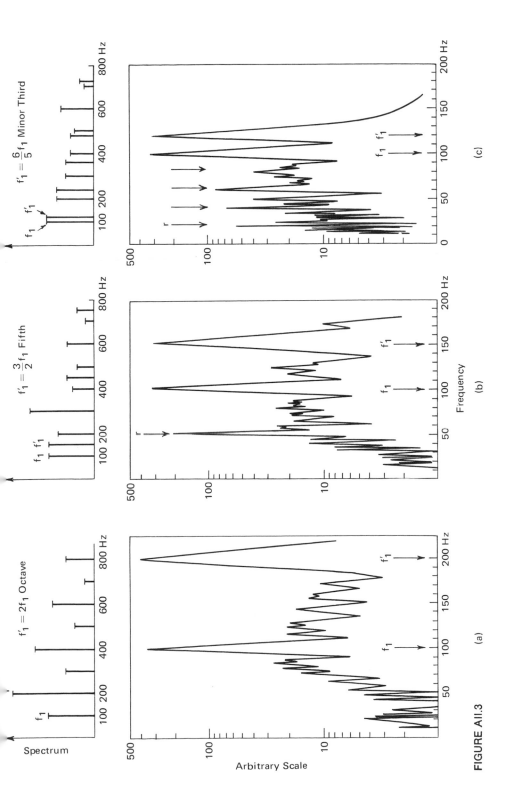

FIGURE AII.3

Appendix III:
Some Remarks on
Teaching Physics and
Psychophysics of Music

It would be unrealistic to make elaborate recommendations on how to organize a truly interdisciplinary course on this subject. The main reason lies in the rather unpredictable composition of the typical student audience signing up for such a course, their widely differing background, and the broad spectrum of interests. Assuming that such a course is open to the whole student body of a university (which it should), it may include four main populations: majors in (1) music; (2) psychology and the biosciences; (3) mass communications and engineering; and (4) physics and mathematics. The single most general difficulty is to make the course equally interesting and easily understandable for everyone. This imposes three over-all requirements: (1) To minimize the use of mathematics—yet to do it in such a way as not to ridicule the presentation in the eyes of the science majors and engineers. (*Hint:* use the course to show science majors explicitly *how to teach science without math!*) (2) To explain everything "from the beginning," be it a topic of physics, psychophysics, or music—yet to do it in such a way as not to appear condescending to the respective "experts." (*Hint:* use the course to show the "experts" explicitly *how to present concise and comprehensive reviews* of topics in their own fields!) (3) To conduct class demonstrations, experiments, and to assign quizzes, essays, and problems that are "meaningful," i.e., conducted in such a way that the student (irrespective of his background) may answer the following question without hesitation: What have I learned by watching the demonstration, by doing this experiment, or by solving this problem? (*Hints:* in the experiments, don't let students watch or make dull measurements for the sake of the measurement—show them *how magnitudes relate to each other in nature*, how they change with respect to each other, and how they are connected through physical cause-and-effect relationships. In the quizzes and problems, don't let them "solve equations"—again, show how a given relationship connects two or more quantities "dynamically" over a whole range of variability, lead them toward *an intuitive "feeling" of quantitative relationships between magnitudes*; show them how mathematical relationships can be used to predict the behavior of a system.)

One serious difficulty is that many music (or other fine arts) majors have an inherent "fear" of scientific rigor, in the sense that they assume *a priori* that "they will not understand." This is exclusively a mental block that can be dispersed successfully with patience, persuasion, and dedication to the individual on part of the teacher. (*Hint:* Convince them that if they are able to balance their checking account each month, they will be able to understand the little bit of math needed in this course!)

The inclusion of psychoacoustics in an introductory course of musical acoustics presents a number of additional challenges to the instructor. First, there is the most obvious one: how to fit everything into the available time. No matter how short or how long that time is, hard decisions will have to be made on which topics should be left out and which should be included. Second, psychoacoustics and the related topics of neuropsychology are subjects perhaps even less familiar to the students of a musical acoustics course than is physics. This makes it necessary to restrict these topics to just a few relevant and interesting ones.

It will help to point out right at the beginning of the course some relevant aspects of psychoacoustics. For instance, alert the student that the recent insights gleaned in music perception *can* be incorporated into the creation of new frontiers in music composition. Point out that many of the fallacies that exist about musical performance *do* have a root in the particular modes of acoustical information-processing in the ear and the brain. Point out that many technical requirements of high-quality electroacoustic equipment are related directly to particular aspects of signal processing in the nervous system. Point out that the understanding of music perception should not only be of interest to musicians but that it can also benefit neuropsychologists in terms of learning about certain brain functions, and that psychologists may obtain from this field quantitative information of relevance to music therapy.

Then, again right at the beginning of the course, pose some fundamental questions of direct relevance to music. Begin with some "ultimate questions," such as: What in music can be explained in terms of physiological properties of the ear and operational modes of the nervous system, and what has emerged "at random" through the development of individual cultures? Point out that music is made up of a succession in time of a certain, very particular class of patterns of acoustical signals, and then ask: Why did music emerge as an important component of human culture if these acoustical patterns are seemingly so irrelevant from a biological survival-value point of view? To what extent is music related to the emergence of language? Then proceed to more specific or narrower questions. Such as: why do we perceive a musical tone, made up of a superposition of many harmonics, as a single whole of one pitch, one loudness, one timbre? Why does the octave play the role of a privileged interval in practically all musical cultures and why do we call the notes differing by octaves with the same name? Why are there scales? Why is there tonal affinity? To what extent are the inharmonicities of musical instruments "imprinted" in the pitch processor of the brain and to what extent does this affect our perception of musical intervals?

A general difficulty is the fact that sophisticated experimental demonstrations concerning this subject involve very expensive equipment. Nevertheless, it is possible to get along with a minimum baseline that only requires equipment that most likely can be borrowed from other courses or departments. This minimum baseline, with which indeed most of the experiments briefly touched upon in this book can be demonstrated, is summarized below.

PSYCHOACOUSTIC EXPERIMENTATION

(1) Two sine-wave generators, a good-quality amplifier, good headphones for everybody, two hi-fi loudspeakers. (2) One oscilloscope for each group of four to six students, if possible, with double-beam display and memory trace. (3) An electronic synthesizer (a portable version is sufficient[1]). With this equipment it is perfectly feasible to demonstrate almost everything that is mentioned on pp. 16, 18, 23, 29, 32, 34, 36–37, 39–40, 87, 136–137, 149–150, and 159. If the class has access to a large pipe organ in a good acoustical environment, additional useful demonstrations can be made (e.g., see pp. 42, 86, 88, 132, 137, and 144). In all these experiments or demonstrations, it should be a rule that *whatever is being sounded should always be simultaneously displayed on the oscilloscope.*

ACOUSTIC EXPERIMENTATION

(1) One "sonometer"[2] per group of four to six students, with stroboscope and appropriate circuitry to conduct the experiments described on pp. 97–98. This setup also provides for experimental discussion of relations (4.2), (4.3), and of bow-

---

[1] See the excellent description of selected experiments and classroom demonstrations given by Hartmann (1975).

[2] A silly name for a single string mounted on a resonance box, with controllable and measurable tension.

ing and plucking mechanisms. (2) A piano is useful for performing the simple demonstrations of pp. 97–99. (3) Single organ pipes (usually available in physics departments) to explore resonance curves of the type of Fig. 4.24, and relations (4.5), (4.6), by using a small speaker of good quality "implanted" in the pipe. (4) Film loops and ripple tanks, also usually available, are extremely useful to demonstrate traveling waves, standing waves, and acoustical "optics" in general.

In addition to all this, it is advisable to assign individual students to conduct special studies and write essays on a musical instrument of their choice, which, of course, will require access to appropriate literature.

To sum up, this is a course that is challenging and fun to teach—the perhaps most interdisciplinary of all that a university can offer at a freshman level. It presents a chance to both teacher and student alike, to let the imagination fly high—within the strict boundaries of science!

References

To find reference in text, look up first author in Index.

Backus, J. 1969. *The Acoustical Foundations of Music.* W. W. Norton and Company, Inc., New York.

Backus, J. 1974. Input impedance curves for the reed woodwind instruments. *J. Acoust. Soc. Amer.* **56**:1266.

Backus, J., and T. C. Hundley, 1971. Harmonic generation in the trumpet. *J. Acoust. Soc. Amer.* **49**:509.

Benade, A. H. 1971. Physics of wind instrument tone and response. *Symposium on Sound and Music, December 1971.* American Association for the Advancement of Science, Washington, D.C.

Benade, A. H. 1973. The physics of brasses. *Sci. Amer.* **229**(1):24.

Benenzon, R. O. 1971. *Musicoterapia y Educación.* Editorial Paidós, Buenos Aires.

Bever, T. G., and R. J. Chiarello. 1974. Cerebral dominance in musicians and nonmusicians. *Science* **185**:537.

Bilsen, F. A., and J. L. Goldstein. 1974. Pitch of dichotically delayed noise and its possible spectral basis. *J. Acoust. Soc. Amer.* **55**:292.

Braitenberg, V. 1974. Thoughts on the cerebral cortex. *J. Theor. Biolog.* **46**:421.

Bredberg, G., H. H. Lindemann, H. W. Ades, R. West, and H. Engstrom, 1970. Scanning electron microscopy of the organ of Corti. *Science* **170**:861.

Brodal, A. 1969. *Neurological Anatomy.* Oxford University Press, New York.

Chistovich, L. A. 1962. Temporal course of speech sound perception. *Proc. 4th Int. Congr. Acoust., Copenhagen.*

Corso, J. F. 1957. Absolute judgments of musical tonality. *J. Acoust. Soc. Amer.* **29**:138.

Culver, C. A. 1956. *Musical Acoustics.* McGraw-Hill Book Company, New York.

Dallos, P., M. C. Billone, J. D. Durrant, C.-Y. Wang, and S. Raynor, 1972. Cochlear inner and outer hair cells: Functional differences. *Science.* **177**:356.

Damaske, P. 1971. Head-related two-channel stereophony with loud-speaker reproduction. *J. Acoust. Soc. Amer.* **50**:1109.

Davis, H. 1962. Advances in the neurophysiology and neuroanatomy of the cochlea. *J. Acoust. Soc. Amer.* **34**:1377.

Eccles, J. C. 1970. *Facing Reality.* Springer-Verlag, New York.

Egan, J. P., and H. W. Hake. 1950. On the masking pattern of a simple auditory stimulus. *J. Acoust. Soc. Amer.* **22**:622.

Flanagan, J. L. 1972. *Speech Analysis, Synthesis and Perception,* 2nd ed. Springer-Verlag, New York.

Fletcher, H., and W. A. Munson. 1933. Loudness, its definition, measurement and calculation. *J. Acoust. Soc. Am.* **5**:82.

Friedlander, F. G. 1953. On the oscillations of a bowed string. *Cambridge Phil. Soc. Proc.* **49**:516.

Gazzaniga, M. S. 1970. *The Bisected Brain.* Meredith Corporation.

Gerstein, G. L., and N. Kiang. 1964. *Exp. Neurol.* **10**:1.

Geschwind, N. 1972. Language and the brain. *Sci. Amer.* **226**(4):76.

Geschwind, N., and W. Levitzky. 1968. Human brain: Left–right asymmetries in the temporal speech region. *Science* **161**:186.

Goldstein, J. L. 1970. Aural combination tones. In *Frequency Analysis and Periodicity Detection in Hearing.* R. Plomp and G. F. Smoorenburg, eds. A. W. Suithoff, Leiden.

Goldstein, J. L. 1973. An optimum processor theory for the central formation of the pitch of complex tones. *J. Acoust. Soc. Amer* **54**:1496.

Gordon, B. 1972. The superior colliculus of the brain. *Sci. Amer.* **227**(6):72.

Hartmann, W. M. 1975. The electronic music synthesizer and the physics of music. *Amer. J. Phys.* **43**:755

Herrington, R. N., and P. Schneidau. 1968. The effect of imagery on the waveshape of the visual evoked response. *Experientia* **24**:1136.

Houtgast, T. 1972. Psychophysical evidence for lateral inhibition in hearing. *J. Acoust. Soc. Amer.* **51**:1885.

Houtsma, A. J. M., and J. L. Goldstein. 1972. Perception of musical intervals: Evidence for the central origin of the pitch of complex tones. *J. Acoust. Soc. Amer.* **51**:520.

Hubel, H. 1971. The visual cortex of the brain. In *Contemporary Psychology. Readings from Scientific American.* W. H. Freeman and Company, San Francisco, California.

Hutchins, C. M., and F. L. Fielding. 1968. Acoustical measurements of violins. *Phys. Today* **21**(7):34.

Jansson, E., N.-E. Molin, and H. Sundin. 1970. Resonances of a violin body studied by hologram interferometry and acoustical methods. *Physica Scripta* **2**:243.

John, E. R. 1972. Switchboard versus statistical theories of learning and memory. *Science* **177**:850.

Keller, J. B. 1953. Bowing of violin strings. *Comm. Pure Appl. Math.* **6**:483.

Kiang, N. Y.-S., T. Watanabe, E. C. Thomas, and L. F. Clark. 1965. *Discharge Patterns of Single Fibers in the Cat's Auditory Nerve.* MIT Press, Cambridge, Massachusetts.

Kimura, D. 1963. Right temporal lobe damage. *Arch. Neurol.* **8**:264.

Kimura, D. 1973. The asymmetry of the human brain. *Sci. Amer.* **228**(3):70.

Klein, W., R. Plomp, and L. C. W. Pols. 1970. Vowel spectra, vowel spaces and vowel identification. *J. Acoust. Soc. Amer.* **48**:999.

Laws, P. 1973. Enternungshören und das Problem der Im-Kopf-Lokalisiertheit von Hörereignissen. *Acustica* **29**:243.

Licklider, J. C. R. 1959. Three auditory theories. In *Psychology: A Study of a Science.* Vol. I. S. Koch, ed. McGraw-Hill Book Company, New York.

Lipps, T. 1905. *Psychologische Studien.* Durr'sche Buchhandlung, Leipzig. Discussed in R. W. Lundin, 1967.

Lundin, R. W. 1967. *An Objective Psychology of Music.* 2nd ed. The Ronald Press Company, New York.

Lynn, P. A., and B. McA. Sayers. 1970. Cochlear innervation, signal processing and their relation to auditory time–intensity effects. *J. Acoust. Soc. Amer.* **47**:525.

Matthews, M. V., and J. Kohut. 1973. Electronic stimulation of violin resonances. *J. Acoust. Soc. Amer.* **53**:1620.

Meyer, M. 1900. Elements of a psychological theory of melody. *Psych. Rev.* **7**:241.

Milner, B. 1967. Brain mechanisms suggested by studies of temporal lobes. In *Brain Mechanisms Underlying Speech and Language.* C. H. Millikan and F. L. Darley, eds., Grune & Stratton, New York and London.

Milner, B., L. Taylor, and R. W. Sperry. 1968. Lateralized suppression of dichotically-presented digits after commissural section in man. *Science* **161**:184.

Molino, J. A. 1973. Pure-tone equal-loudness contours for standard tones of different frequencies. *Percept. Psychophys.* **14**:1.

Molino, J. A. 1974. Psychophysical verification of predicted interaural differences in localizing distant sound sources. *J. Acoust. Soc. Amer.* **55**:139.

Moore, B. C. J. 1973. Frequency difference limens for short-duration tones. *J. Acoust. Soc. Amer.* **54**:610.

Papçun, G., S. Krashen, D. Terbeek, R. Remington, and R. Harshman. 1974. Is the left hemisphere specialized for speech, language and/or something else? *J. Acoust. Soc. Amer.* **55**:319.

Patterson, B. 1974. Musical dynamics. *Sci. Amer.* **231**(5): 78.

Penfield, W., and L. Roberts. 1959. *Speech and Brain Mechanisms.* Princeton University Press, Princeton, New Jersey.

Plomp, R. 1964. The ear as a frequency analyzer. *J. Acoust. Soc. Amer.* **36**:1628.

Plomp, R. 1965. Detectability threshold for combination tones. *J. Acoust. Soc. Amer.* **37**:1110.

Plomp, R. 1967. Beats of mistuned consonances. *J. Acoust. Soc. Amer.* **42**:462.

Plomp, R. 1967a. Pitch of complex tones. *J. Acoust, Soc. Amer.* **41**:1526.

Plomp, R. 1970. Timbre as a multidimensional attribute of complex tones. In *Frequency Analysis and Periodicity Detection in Hearing.* R. Plomp and F. G. Smoorenburg, eds. A. W. Suithoff, Leiden.

Plomp, R., and M. A. Bouman. 1959. Relation between hearing threshold and duration for tone pulses. *J. Acoust. Soc. Amer.* **31**:749.

Plomp, R., and W. J. M. Levelt. 1965. Tonal consonance and critical bandwith. *J. Acoust. Soc. Amer.* **38**:548.

Plomp, R., and G. F. Smoorenburg, editors. 1970. *Frequency Analysis and Periodicity Detection in Hearing.* A. W. Suithoff, Leiden.

Plomp, R., and H. J. M. Steeneken. 1971. Pitch versus timbre. *Proc. 7th Int. Congr. Acoust., Budapest.*

Plomp, R., and H. J. M. Steeneken. 1973. Place dependence of timbre in reverberant sound fields. *Acustica* **28**:50.,

Pribram, K. H. 1971. *Languages of the Brain.* Prentice-Hall, Inc., Englewood Cliffs, New Jersey.

Rakowski, A. 1971. Pitch discrimination at the threshold of hearing. *Proc. 7th Int. Congr. Acoust. Budapest.* **3**:373.

Rakowski, A. 1972. Direct comparison of absolute and relative pitch. *Proc. Symp. Hearing Theory, IPO Eindhoven.*

Ratliff, F. 1972. Contour and contrast. *Sci. Amer.* **226**(6):91.

Reinicke, W., and L. Cremer. 1970. Application of holographic interferometry to vibrations of the bodies of string instruments. *J. Acoust. Soc. Amer.* **48**:988.

Rhode, W. S., and L. Robles. 1974. Evidence from Mössbauer experiments for nonlinear vibration in the cochlea. *J. Acoust. Soc. Amer.* **55**:588.

Rigden, J. S. 1974. Variations of sound intensity for mistuned consonances. *J. Acoust. Soc. Amer.* **55**:1095.

Ritsma, R. J. 1967. Frequencies dominant in the perception of the pitch of complex sounds. *J. Acoust. Soc. Amer.* **42**:191.

Roederer, J. G. 1972. La musique, l'oreille et le cerveau. *Découverte* **41** (May).

Roederer, J. G. 1974. Auditory processing in the nervous system. In: *Physics and Mathematics of the Nervous System.* M. Conrad, W. Güttinger, and M. Dal Cin, eds. Springer-Verlag, Berlin, Heidelberg, New York.

Rose, J. E., J. F. Brugge, D. J. Anderson, and J. E. Hind. 1969. Some possible neural correlates of combination tones. *J. Neurophys.* **32**:402.

Saunders, F. A. 1946. The mechanical action of instruments of the violin family. *J. Acoust. Soc. Amer.* **17**:169.

Scheid, P., and Eccles, J. C. 1975. Music and speech: Artistic functions of the human brain. In *Psychology of Music.* (In press.)

Schelleng, J. C. 1973. The bowed string and the player. *J. Acoust. Soc. Amer.* **53**:26.

Siebert, W. M. 1970. Frequency discrimination in the auditory system: Place or periodicity mechanisms? *Proc. IEEE* **58**:723.

Simmons, F. B. (1970). Monaural processing. In *Foundations of Modern Auditory Theory.* J. V. Tobias, ed. Academic Press, New York.

Small, A. M. 1970. Periodicity pitch. In *Foundations of Modern Auditory Theory.* J. V. Tobias, ed. Academic Press, New York.

Smoorenburg, G. F. 1970. Pitch perception of two-frequency stimuli. *J. Acoust. Soc. Amer.* **48**:924.

Smoorenburg, G. F. 1972. Audibility region of combination tones. *J. Acoust. Soc. Amer.* **52**:603.

Sokolich, W. G., and J. J. Zwislocki. 1974. Evidence for phase oppositions between inner and outer hair cells. *J. Acoust. Soc. Amer.* **55**:466.

Sommerhoff, G. 1974. *Logic of the Living Brain.* John Wiley and Sons, London, New York, Sydney, Toronto.

Spoendlin, H. 1970. Structural basis of peripheral frequency analysis. In *Frequency Analysis and Periodicity Detection in Hearing* R. Plomp and G. F. Smoorenburg, eds. A. W. Suithoff, Leiden.

Stevens, S. S. 1955. Measurement of loudness. *J. Acoust. Soc. Amer.* **27**:815.

Stevens, S. S. 1970. Neural events and the psychophysical law *Science* **170**:1043

Stevens, S. S., J. Volkmann, and E. B. Newman. 1937. A scale for the measurement of psychological magnitude pitch. *J. Acoust. Soc. Amer.* **8**:185.

Terhardt, E. 1971. Pitch shifts of harmonics, an explanation of the octave enlargement phenomenon. *Proc. 7th Int. Congr. Acoust.* Budapest. **3**:621.

Terhardt, E. 1972. Zur Tonhöhenwahrnehmung von Klängen, I, II. *Acustica* **26**:173, 187.

Terhardt, E. 1974. Pitch, consonance and harmony. *J. Acoust. Soc. Amer.* **55**:1061.

Terhardt, E., and H. Fastl. 1971. Zum Einfluss von Störtönen und Störgeräuschen auf die Tonhöhe von Sinustönen. *Acustica* **25**:53.

Terhardt, E. and M. Zick. 1975. Evaluation of the tempered tone scale in normal, stretched and contracted intonation. *Acustica* **32**:268.

Tobias, J. V., editor. 1970. *Foundations of Modern Auditory Theory* Academic Press, New York.

van Noorden, L.A.P.S. 1975. *Temporal coherence in the perception of tone sequences* (with a phonographic demonstration record) Institute for Perception Research, Eindhoven.

von Békésy, G. 1960. *Experiments in Hearing.* McGraw-Hill Book Company, New York.

von Helmholtz, H. 1863. *On the Sensations of Tone as a Physiological Basis for the Theory of Music.* English translation, 1954 Dover Publications, New York.

Walliser, K. 1969. Über die Abhängigkeiten der Tonhöhenempfindung von Sinustönen von Schallpegel, von überlagertem drosselnden Störschall und von der Darbietungsdauer. *Acustica* **21**:211.

Ward, W. D. 1970. Musical perception. In *Foundations of Modern Auditory Theory.* J. V. Tobias, ed. Academic Press, New York

Wightman, F. L. 1973. The pattern-transformation model of pitch. *J. Acoust. Soc. Amer.* **54**:407.

Wightman, F. L., and D. M. Green. 1974. The perception of pitch. *American Scientist* **62**:208.

Whitfield, I. C. 1967. *The Auditory Pathway.* Edward Arnold Ltd., London.

Zwicker, E. G. Flottorp, and S. S. Stevens. 1957. Critical bandwidth in loudness summation. *J. Acoust. Soc. Amer.* **29**:548.

Zwislocki, J. J. 1965. Analysis of some auditory characteristics. In *Handbook of Mathematical Psychology.* R. D. Luce, R. R. Bush, and E. Galanter, eds. Wiley, New York.

Zwislocki, J. J. 1969. Temporal summation of loudness: An analysis. *J. Acoust. Soc. Amer.* **46**:431.

Zwislocki, J. J., and W. G. Sokolich. 1973. Velocity and displacement responses in auditory-nerve fibers. *Science* **182**:64.

Index

Page numbers in bold face locate main definition or principal explanation.

A

Absorption coefficient, **131**
Action potential, **46**. *See also* Neural impulse
Air column, open, 115
Air column, stopped, 117
Amplitude, **17**
Antinodes, **75**, 77, 95, 101, 117
Auditory pathway (or channel), **57**, 167
Aural harmonics, **36**
Autocorrelation (of neural signals), 51, 56
Axon, 45, **179**

B

Backus: (1969), *viii*; (1974), 123
Backus and Hundley, 126
Basilar membrane, **19**, 27, 31, 45, 48, 134, 168
Beats (first-order), **27**, 32, 39, 147
Beats (second-order), **37**, 39, 50
"Beautiful" instruments, 142
Benade: (1971), 122, 123, 126, 127; (1973), 128
Benenzon, 165
Beaver and Chiarello, 169
Bilsen and Goldstein, 57
Body of a violin, 110
Bowing mechanism, **105**, 171
Braitenberg, 163
Brass instruments. 121, 126
Bredberg et al., 20, 21
Brodal, 57

C

Categorical mode (of pitch perception), **160**
Cent (unit of frequency ratio), **157**
Center frequency, **30**
Central pitch processor, **54**, 135, 138, 150, 159, 162, **178**
Cerebral hemispheres, 12, 146, **165**, 169
Chistovich, 163
Chorus effect, 145
Chroma, 151
Clarinet-type instrument, 118, 122, 124
Cochlea, **19**
Cochlear duct. *See* Cochlea
"Cocktail party effect," **146**
Coding (in the auditory system), **21**, 49, 51, 56, 90, 135, 138, 150, 182
Combination tones, **34**, 36, 177
Complex tone, **44**, 108, 134, 144,, 149, 180
Consonance, 147, 150, 152, 182
Contrast enhancement. *See* Sharpening
Corpus callosum, **60**, 166
Corso, 160
Cortex (cerebral), **6**, 59, 139, 146
Critical band, **28**, 30, 33, **85**, 109, 134, 149
Crosscorrelation (of neural impulses), **52**, 59
Culver, 108
Cycle, 15
Cycles per second. *See* Hertz

D

Dallos et al., 45
Damaske, 133
Damped oscillation, **64**, 102
Davis, 45
db. *See* Decibel
Decay. *See* Damped oscillation
Decibel, **80**
Dendrite, **45**
Difference tone, **34**
Diffraction, **132**
Directionality (of sound). *See* Localization
Dissonance, 147
Dominance (of a tone in a sequence). *See* Finality
Dominant hemisphere, cerebral, **166**, 168
Duration (effect on loudness), 88

E

Eardrum, **19**, 63
Eccles, 45, 141
Edge tone, **120**
Egan and Hake, 87
Elastic waves, **61**, 71
Emotionality (in music perception), **164**
Energy, **63**

Engram, **141**
Enharmonic equivalent, **155**
Excitation mechanism, **1**, 99, **119**
Excitatory synapse, **47**
Experiments, acoustic, **186**
Experiments, psychoacoustic, **186**

F

Feature detection, 48, 59, 139
Finality (tonal), **162**
Fingerholes (effect of), **126**, 128
Firing pattern (neural), **140**
Firing rate (neural), **48**, 90
Fission (of melodies), **162**
Flanagan, 20, 89, 128, 134
Fletcher and Munson, 82, 83
Flute-type instruments, 116, 122
Force, **62**
Formants, **114**, 128
Fourier analysis, **107**. *See also* Spectrum
Frequency, **19**, 69
Frequency discrimination, **27**, 30
Frequency, fundamental, **40**, 95, 116, 118, 124, 148
Frequency resolution, **23**
Frequency standard, **160**
Friction: Static, **171**; Dynamic, **172**
Friedlander, 105
Fundamental tracking, **44**. *See also* Subjective pitch

G

Gerstein and Kiang, 140
Geschwind, 166
Geschwind and Levitzky, 170
Gestalt (musical), 5, 135, 143, 162, 168
Goldstein: (1970), 34, 35; (1973), 54, 55, 175
Gordon, 146
Gazzaniga, 166

H

Hair cells, **20**, 45, 90
Half tone. *See* Semitone
Harmonics, **43**, 95, 98, 106, 118, 125, 133, 144, 147
Hartmann, 185
Hemispheres. *See* Cerebral hemispheres
Herrington and Schneidau, 141
Hertz (unit of frequency), **19**
Hi-fi systems, 36, 109, 133
Holography, **110**
Houtgast, 91
Houtsma and Goldstein, 41, 42, 55, 137, 150, 161
Hubel, 92, 139
Hutchins and Fielding, 113
Hz. *See* Hertz

I

IL. *See* Sound intensity level
Identification (of musical instruments). *See* Recognition
Image (auditory), **140**, 142, 168
Information, neural. *See* Coding
Inharmonic (vibration, mode, tone), **97**, 119, 123, 177
Inhibitory synapse, **47**, 181
Inner ear. *See* Cochlea
Input impedance, 122
Intensity, **73**, 130
Intensity level. *See* Sound intensity level
Interval stretching, 91, 138, 151, **159**
Intervals, musical, 39, **148**, 158

J

Jansson et al., 110, 111
jnd of frequency, **23**
jnd of intensity, **79**
John, 140, 141, 163
Joule (unit of energy), **64**
Just noticeable difference (*jnd*), **10**, 23
Just scale, **154**, 158

K

Keller, 105
Key color, **160**
Kiang et al., 49
Kimura: (1963), 166; (1973), 167
Klein et al., 137

L

L. *See* Loudness, subjective
LL. *See* loudness level
Laws, 133
Learning (as a neural process), 56, **139**, 150, **180**
Licklider, 51, 52, 53
Limbic system, **164**
Linear superposition. *See* Superposition of vibrations
Lipps, 162
Localization (of sound), **52**
Logarithms, **79**
Loudness, **3**, 19, 78, 89, 105
Loudness level, **84**
Loudness, subjective, **85**, 90, 136, 142
Loudness summation, 86
Lundin, *viii*
Lynn and Sayers, 92

M

ML. *See* Masking level
Magnitude, psychophysical, **10**
Major triad. *See* Triad

Masking. *See* Masking level
Masking level, **81**, 87
Matthews and Kohut, 142
Melody fission. *See* Fission
Melody (melodic line), 137, 145, 162
Memory, **142**, 163, 165. *See also* Remembering
Meyer, 162
Milner, 166
Milner et al., 60
Minor hemisphere, **166**, 169
Missing fundamental, **42**, 55, 180. *See also* Periodicity pitch
Modes (of vibration), **96**, 98, 110
Molino: (1973), 82; (1974), 52
Moore, 56
Mössbauer effect, **22**
Motion, 13, **14**
Motion, harmonic, **17**, 64
Motion, periodic. *See* Vibration
Musicality, 170

N

Nervous system (auditory). *See* Auditory pathway
Neural impulse (*also* pulse), **45**, 49, 90
Neural networks (neural models), 141, **178**
Neural signal. *See* Neural impulse
Neuron, **45**, 139, 179
Newton (unit of force), **62**
Nodes, **75**, 77, 94, 115
Noise, **153**

O

Oboe-type instrument, 122, 124
Octave, 22, 147, **152**, 182
Organ, 42, 86, 125, 133, 137
Organ pipes, **125**
Oscillation. *See* Vibration
Oscilloscope, **16**
Overblow, **125**
Overtones, **95**. *See also* Harmonics

P

Papçun et al., 167
Patterson, 78
Penfield and Roberts, 166
Period, **17**
Periodicity pitch, **42**, **54**. *See also* Pitch, subjective
Phase, **18**, 26, 38, 132, 136
Phase difference. *See* Phase
Phon (unit of loudness level), **84**
Phrasing, **88**
Physics, **7**
Piano, experiments with, **97**, 98
Pipe (open, stopped). *See* Air column

Pitch, **3**, 19, 21, 23, 57, 91, 124, 135. *See also* Pitch, subjective
Pitch, absolute, **160**
Pitch, ambiguous (multiple), 55, 135, **177**, 180
Pitch discrimination (of complex tones), **152**
Pitch matching experiments, **28**
Pitch processor. *See* Central pitch processor
Pitch, spectral, **42**, 179
Pitch, subjective, **42**, **54**, 135, 138, 160, 176. *See also* Pitch,
    Periodicity pitch, Central pitch processor
Place theory (of hearing, pitch), **24**, 44, 50, 57
Planum temporale, **170**
Plomp: (1964), 134, 138; (1965), 34; (1967), 39, 40; (1967a), 42, 44;
    (1970), 136
Plomp and Bowman, 88
Plomp and Levelt, 148, 149
Plomp and Smoorenburg, *viii*
Plomp and Steeneken: (1971), 136; (1973), 132
Power, **65**, 74
Power spectrum. *See* Spectrum
Precedence effect, **129**, 133, 146
Pressure, **63**
Pressure oscillations. *See* Sound waves
Pressure variation, **63**, 71, 80
Pribram, **140**
Propagation (of sound), **61**
Psychophysics, **8**
Pure tone, **19**, 50
Pythagorean scale, **155**, 158

Q

Quality of sound. *See* Timbre
Quantum physics, 8

R

Rakowski: (1971). 23; (1972), 161
Ratliff, 91
Read-out mechanism (in the brain), 9, **140**
Recognition (of a musical instrument), 115, **141**
Reed, **120**, 124, 138
Reflection (of sound), **72**, 129
Register (low, middle, top, of an instrument), **124**, **125**
Reinicke and Cremer, 110
Remembering (as a neural process), **140**, 142
Repetition rate, **40**, 49, 107, 135. *See also* fundamental frequency
Residue tone. *See* Missing fundamental
Resonance curve, **113**, 122, 126, 142
Resonance frequency, **112**. *See also* Resonance curve
Resonance peaks. *See* Resonance curve
Resonance region (on the basilar membrane) **21**, 135, 144, 168, 175
Resonator, **102**, 109
Response curve. *See* Resonance curve
Return, sense of. *See* Finality
Reverberation, **130**, 146
Reverberation time, 130

Rhode and Robles, 23
Rhythm, 165
Rigden, 39
Ritsma, 44
Roederer: (1972), 163; (1974), 141
Rose et al., 51
Roughness (of two pure tones), **28**, 147

S

SPL. *See* Sound pressure level
Saunders, 142
Scale, musical, **153**, 158, 161
Scheid and Eccles, 165, 169, 170
Schelleng, 105
Semitone, **154**, 155, 157
Sharpening, 32, **91**
Siebert, 50
Simmons, 139
Sinusoidal motion. *See* Motion, harmonic
Small, 42
Smoorenburg: (1970), 177; (1972), 36
Sokolich and Zwislocki, 47
Sommerhoff, *viii*
Sone (unit of loudness), 85
Sound intensity level, **80**
Sound pressure level, **80**
Sound waves. *See* Waves
Spectrum, **108**, 114, 122, 128, 132, 153
Speech center (of the brain), **166**
Spike. *See* Neural impulse
Spoendlin, 20, 45
Stereo perception (of sound). *See* Sound localization
Stevens: (1955), 84; (1970), 84, 85
Stevens et al., 24
Stretched intervals. *See* Interval stretching
Strings, **92**, 98, 171
Superior colliculus, 59, **146**
Superposition (of vibrations, waves), **25**, 74, 100, 143
Synapse, **46**, 179
Synthesis (of tones), **107**, 133

T

Teaching, **184**
Temperature (effect on frequency), **68**, 116, 118
Tempered scale, **156**, 158
Terhardt: (1971), 138; (1972), 54, 55, 150; (1974), 54, 55, 56, 150,
    151, 178, 179
Terhardt and Fastl, 91
Terhard and Zick, 159
Thinking (as a neural mechanism), 141, **168**
Timbre, **4**, 98, 105, 114, 125, **136**, 142, 150
Timbre discrimination, **144**
Tobias, *viii*
Tonal meaning, **150**

Tonality, 156, 158, 163
Touch (in piano playing), **104**
Transients, 4, **92**, 109, 132, 144
Triad, 153, 154
Tuning, 33

V

van Noorden, 163
Velocity, **15**, 67, 68
Velocity (of waves), 67, 69
Vibration, **15**
Vibration pattern, **15**, 38, 48, 51
Vibrato, 142, 145
Virtual pitch. *See* Subjective pitch
Volleys (of nerve impulses), 50
von Békésy, 21, 22
von Helmholtz, *vii*

W

Walliser, 91
Ward, 159
Watt (unit of power), **65**
Wavelength, **68**, 70, 72, 94
Waves, logitudinal, **67**, 70, 116
Waves (sound), 61, **63**
Waves, standing, **75**, 94, 116, 133
Waves, transverse, **67**, 94
Whole tone, **154**, 155
Woodwind instruments, 121, 125
Work, **63**
Wightman, 54, 56, 182
Wightman and Green, 53
Whitfield, 57

Z

Zwicker et al., 23, 24, 30, 85
Zwislocki: (1965), 33; (1969), 89
Zwislocki and Sokolich, 48